Origins of Life

A cosmic perspective

Origins of Life

A cosmic perspective

Douglas Whittet
New York Center for Astrobiology
Department of Physics, Applied Physics and Astronomy
Rensselaer Polytechnic Institute

Morgan & Claypool Publishers

ISBN   978-1-6817-4676-0 (ebook)
ISBN   978-1-6817-4677-7 (print)
ISBN   978-1-6817-4678-4 (mobi)

DOI   10.1088/978-1-6817-4676-0

Version: 20171101

IOP Concise Physics
ISSN 2053-2571 (online)
ISSN 2054-7307 (print)

A Morgan & Claypool publication as part of IOP Concise Physics
Published by Morgan & Claypool Publishers, 1210 Fifth Avenue, Suite 250, San Rafael, CA, 94901, USA

IOP Publishing, Temple Circus, Temple Way, Bristol BS1 6HG, UK

For Polly, Clair and James, and in memory of my parents

Contents

Preface	**xi**
Acknowledgements	**xii**
Author biography	**xiii**

1	**Introduction**	**1-1**
1.1	What is life?	1-2
1.2	Could life be silicon-based?	1-3
1.3	Time constraints	1-5
1.4	Scenarios for the origin of life on Earth	1-7
	1.4.1 The prebiotic world	1-7
	1.4.2 The protobiotic world	1-9
	1.4.3 Organization	1-11
	1.4.4 Life from elsewhere?	1-11
1.5	Searching for life beyond Earth	1-12
	Questions and discussion topics	1-13
	References and further reading	1-14

2	**Cosmic synthesis of the biologically important chemical elements**	**2-1**
2.1	Elements essential to life	2-1
2.2	Origin of the elements	2-3
	2.2.1 A primer on nuclear fusion	2-3
	2.2.2 Element production in the early Universe	2-5
	2.2.3 Element production in stars	2-5
	2.2.4 Element production in supernovae	2-9
2.3	Distributing the products of nucleosynthesis	2-10
	2.3.1 Low and high mass stars: a cosmic case of 'The tortoise and the hare'	2-11
	2.3.2 Carbon stars	2-11
	2.3.3 Metallicity over time	2-13
	Questions and discussion topics	2-14
	References and further reading	2-15

3 Molecules in space: from interstellar clouds to protoplanetary disks — 3-1

3.1 Chemistry in the interstellar medium — 3-2
   3.1.1 Interstellar environments — 3-2
   3.1.2 Surface chemistry — 3-3
   3.1.3 Gas-phase chemistry — 3-4
   3.1.4 Deuterium fractionation as a diagnostic of interstellar chemistry — 3-5
3.2 The rise of molecular complexity — 3-6
   3.2.1 Organic interstellar molecules and the search for glycine — 3-7
   3.2.2 Hydrogenation versus oxidation: a vital branching point for surface chemistry — 3-7
3.3 Protostars and the chemical heritage of protoplanetary disks — 3-8
   3.3.1 A brief primer on star formation — 3-8
   3.3.2 Chemical evolution in a protostellar envelope — 3-11
   3.3.3 The significance of the birth environment — 3-12
   Questions and discussion topics — 3-14
   References and further reading — 3-14

4 The origin and evolution of our solar system — 4-1

4.1 The Sun's birth environment — 4-1
4.2 The solar nebula and the origin of the planets — 4-3
   4.2.1 Overview — 4-3
   4.2.2 Frost and soot lines — 4-3
   4.2.3 Accretion: from dust to planets — 4-5
   4.2.4 Origins of the giant planets — 4-6
4.3 Time capsules from the early Solar System — 4-8
   4.3.1 Comets and asteroids — 4-8
   4.3.2 Meteorites — 4-9
4.4 The evolution of habitability — 4-11
   4.4.1 The circumstellar habitable zone — 4-12
   4.4.2 Energy from within — 4-14
   Questions and discussion topics — 4-15
   References and further reading — 4-15

5 The early Earth: forging an environment for life — 5-1

5.1 The Earth–Moon system — 5-1
5.2 Cosmic impacts — 5-3

	5.2.1	Significance	5-3
	5.2.2	Cratering chronology	5-4
	5.2.3	Late heavy bombardment	5-6
5.3	Emergence of the atmosphere and hydrosphere		5-7
	5.3.1	The message in the rocks	5-8
	5.3.2	Origin of the oceans	5-11
5.4	Toward a prebiotic world		5-13
	5.4.1	Organics from space	5-14
	5.4.2	Geochemical processes	5-15
	Questions and discussion topics		5-16
	References and further reading		5-16

6   The origin of terrestrial life: converging on a paradigm   6-1

6.1	Constructing life's framework		6-2
	6.1.1	The RNA world hypothesis	6-2
	6.1.2	Chiral selectivity	6-4
	6.1.3	Metabolism first?	6-5
	6.1.4	Organization and the emergence of cells	6-6
6.2	Tracing life's ancestry		6-7
	6.2.1	The phylogenetic tree	6-7
	6.2.2	Life's early imprint	6-10
6.3	The transition to life		6-11
	Questions and discussion topics		6-13
	References and further reading		6-13

7   The search for life on Mars   7-1

7.1	Evolving conditions and habitability		7-2
	7.1.1	Mars and Earth compared	7-2
	7.1.2	Interpreting the topography	7-3
	7.1.3	Evolution of the Martian climate	7-6
7.2	Searching for biosignatures		7-8
	7.2.1	The Viking biology experiments	7-8
	7.2.2	Martian meteorites	7-9
	7.2.3	Atmospheric methane	7-11
	7.2.4	Searching the sediments	7-12
7.3	Future prospects		7-13
	Questions and discussion topics		7-14
	References and further reading		7-15

8	**Icy worlds as potential hosts for life**	**8-1**
8.1	The moons of Jupiter	8-1
	8.1.1 Europa	8-3
	8.1.2 Ganymede	8-4
8.2	Saturn and beyond	8-6
	8.2.1 Enceladus	8-6
	8.2.2 Titan	8-8
	8.2.3 Triton, Pluto and Charon	8-9
8.3	Prospects for life	8-10
	8.3.1 Hydrothermal systems	8-10
	8.3.2 Subglacial lakes as analog environments	8-11
	8.3.3 Alternatives to water?	8-13
	8.3.4 Future exploration	8-13
	Questions and discussion topics	8-14
	References and further reading	8-15
9	**The search for life beyond our solar system**	**9-1**
9.1	Exoplanetary systems: characteristics and habitability	9-1
	9.1.1 Detection	9-1
	9.1.2 Metallicities of host stars	9-4
	9.1.3 Circumstellar habitable zones	9-5
	9.1.4 Exoplanetary systems compared	9-7
	9.1.5 Super-earths	9-9
	9.1.6 The galactic habitable zone	9-10
9.2	The search for spectroscopic biosignatures	9-11
	9.2.1 The Earth as a template	9-11
	9.2.2 Atmospheric methane and oxygen	9-12
	9.2.3 The red edge	9-13
9.3	The search for extraterrestrial intelligence (SETI)	9-15
	9.3.1 Rationale and brief history	9-15
	9.3.2 The Drake equation	9-16
	9.3.3 Where is everyone?	9-17
9.4	What if we succeed?	9-18
	9.4.1 Dissemination	9-18
	9.4.2 Implications	9-19
	Questions and discussion topics	9-20
	References and further reading	9-21

Preface

How did life on Earth begin? Does it exist elsewhere in the Universe? These are amongst the most fundamental and compelling questions in all science. In its 125th anniversary edition, published in 2005, the journal *Science* included 'How and where did life on Earth arise?' in a list of compelling questions that scientists should have a good shot at answering, or at least know how to go about answering, over the next 25 years. Progress toward this goal depends on research that spans the traditional science disciplines, from mathematics, physics and astronomy to planetary and earth sciences, chemistry, and biology. The advance of science over past centuries has led to an understanding that the basic laws and concepts of mathematics, physics and chemistry, developed for the most part to describe phenomena on our home planet, are to the best of our knowledge universal: we can use them to predict the flightpath of a space mission, to identify molecules in distant galaxies, and to measure the expansion of the Universe. But is there a universal biology? Interdisciplinary research that attempts to answer this question is at the heart of the field now known as astrobiology—the study of the origins, distribution and future of life in the Universe.

This vibrant field is a natural choice for university science courses, providing opportunities not only to engage students' interest and imagination but also to illustrate the advantages of an interdisciplinary approach to problem solving. This book is based on a course of the same title, taught by the author as part of the science curriculum at Rensselaer Polytechnic Institute. The course provides a grounding in the concepts, methods and theories of astrobiology and origins of life research, and presents a summary of the latest findings. The plural 'Origins' in the title is significant. Inevitably, the focus is on the origin of life *on Earth*, the only confirmed example we have of its existence in the Universe, but with an emphasis on the broader context: insight into the environments and processes that gave birth to life on our planet will naturally inform our assessment of the probability that it has arisen (or will arise) elsewhere.

The text is intended to be suitable for mid- and upper-level undergraduates and beginning graduate students, and more generally as an introduction and overview for researchers and general readers seeking to follow current developments in this interdisciplinary field. The reader is assumed to have a basic grounding in the relevant sciences, but prior specialized knowledge is not required. Each chapter concludes with a list of questions and discussion topics, and suggestions for further reading. Some questions can be answered with reference to material in the text, but others require further reading and some have no known answers! The intention is to encourage the reader to go beyond basic concepts, to explore topics in greater depth, and, in a classroom setting, to engage in lively discussions with class members.

Acknowledgements

My thanks are due, first and foremost, to my family: to my late parents Doreen and Douglas Whittet, for their love and support throughout my formative years and beyond, to Polly, my wife and dearest friend, and to my children, Clair and James. I am so proud of you all.

Many have contributed to my development as a scientist, as I evolved from a 'gung ho' astronomer to an aspiring astrobiologist. The process began as long ago as 1971 when, on a whim, I purchased a book entitled *'Planets and Life'* by Peter Sneath—now long out of print but clearly ahead of its time! A motivation for writing a book of my own on this topic is the thought that it might inspire others in the same way that his did for me. I also owe a debt of gratitude to the late Jim Ferris, my predecessor as leader of a NASA-supported research and training center at Rensselaer Polytechnic Institute: for many years Jim carried the banner for Origins of Life research, long before it evolved into Astrobiology and became a popular main-stream field of study.

To my colleagues, friends and co-conspirators in the New York Center for Astrobiology—especially Bruce Watson, Karyn Rogers and Linda McGown—I am so grateful to have had the opportunity to work with you, to learn from you, and to have had so much fun along the way. I also learned much during my time of service on the Executive Council of the NASA Astrobiology Institute, chaired by Carl Pilcher—Carl, you were my perfect role model! And to the many students who have taken the course on which this book is based over the years—I have learned from you as well! Your excitement and enthusiasm amplified mine, and your feedback improved the course and hence, I hope, this text. You may recognize many topics we debated in class in the end-of-chapter sets of questions and discussion topics.

It is a pleasure to acknowledge Joel Claypool, Jeanine Burke, Chris Benson, Melanie Carlson and others in the editorial and production teams at Morgan & Claypool and IoP Publishing, for their patience, encouragement and expert guidance. I am also indebted to all who gave permission to reproduce images included in the figures; these are acknowledged individually in the appropriate caption.

Author biography

Doug Whittet

Doug Whittet is Emeritus Professor of Physics at Rensselaer Polytechnic Institute. He obtained his bachelor and doctoral degrees in Physics and Astronomy, respectively, from the University of St Andrews, and was Senior Lecturer and Professor of Astrophysics at the University of Central Lancashire prior to joining the Rensselaer faculty in 1991. He served as Director of the New York Center for Astrobiology from 2008 to 2016. His research interests include the physics and chemistry of the interstellar medium, with a focus on the chemical inventories of preplanetary matter in the disks and envelopes of newly-formed stars. In the course of his research, he was a principal investigator of observing programs with the Hubble Space Telescope, the Spitzer Space Telescope and the Infrared Space Observatory. He also has a strong interest in undergraduate education and curriculum development in astronomy and related fields, a product of which is the course on which this book is based. *Origins of Life: A Cosmic Perspective* has been part of the science curriculum at Rensselaer Polytechnic Institute since 1995—although it has, like the field itself, evolved dramatically over the years!

IOP Concise Physics

Origins of Life
A cosmic perspective
Douglas Whittet

Chapter 1

Introduction

How life formed, and how it is distributed in the Universe, are distinct but closely related questions. For many years, scientists have considered the possibility that life might have arisen on Earth from non-living precursors in some suitable environment, such as the 'warm little pond' envisioned by Charles Darwin or the hydrothermal vents and fissures in the Earth's crust and ocean floors that are considered likely candidates today. Laboratory experiments have shown that it is possible to create amino acids and other prebiotic molecules from simple precursors such as H_2O, NH_3 and CH_4 in the presence of an energy source. Such experiments provide an example of the 'bottom-up' approach to research on life's origins: tracing molecular evolution from simple to complex. This is balanced by phylogenetic ('top-down') research that attempts to trace evolution backwards from extant life to its earliest microbial ancestors, and thereby to identify the conditions in which it arose. The bottom-up and top-down approaches are complementary, with the obvious goal that they should ultimately converge. The emergence of life on our planet seems certain to have been inextricably linked with environmental conditions and resources available on the early Earth, and these were determined by a combination of internal and external factors. Understanding how and under what conditions life arose on Earth would, in turn, lead to a better assessment of the probability that it exists elsewhere, within or beyond our solar system. This is the cosmic perspective.

The environments and materials available on the prebiotic Earth some 4 billion years (4 Ga) ago were strongly influenced by several external factors, including the Sun's luminosity, the influx of meteoritic and cometary material, the frequency of potentially planet-sterilizing impacts, and the tidal forces arising from the proximity of a large moon. Somewhat similar conditions must have existed on Mars at this time, when its giant volcanoes were presumably active, and its surface temperature and atmospheric pressure were evidently sufficient to support the presence of liquid water. Research on the origin of terrestrial life is therefore highly relevant to the quest to discover whether Mars has (or ever had) life, and provides guidance on the design of missions to search for it. More generally, studies of terrestrial extremophiles—organisms that thrive in extreme environments—provide insight into the potential limits to life in alien environments, including not only Mars but also icy worlds such as Europa and Titan in the outer solar system.

Looking beyond our solar system, advances in observational astronomy have enabled the discovery of multitudes of exoplanets—planets orbiting other stars in our Galaxy—some of which seem likely candidates as hosts for life. As our observational capabilities continue to develop, we are presented with a realistic opportunity to detect spectroscopic signatures that could confirm the existence of microbial life on such distant worlds. Thus, the search for life in the Universe is no longer limited to SETI (the search for extraterrestrial intelligence). Active since the mid-20th century, SETI attempts to detect electromagnetic signals such as radio waves emitted—intentionally or otherwise—by advanced civilizations. Our own planet has hosted microbial life for at least 3.5 Ga (more than 75% of its history; see section 1.3), compared with less than a century for the technology needed for electromagnetic communication: if our case is typical, it would be logical to suppose that a search for microbial biosignatures is more likely to be successful.

The following sections of this chapter review some basic facts, questions and concepts that underpin research in astrobiology. We begin with the fundamental and perplexing issue of how life should be defined (section 1.1) and whether life based on other elements such as silicon could be a feasible alternative to carbon-based life (section 1.2). Constraints on the timing of life's origin on Earth and the processes that may have led to it are assessed in sections 1.3 and 1.4, respectively. Strategies for detecting life beyond our home planet are summarized in the final section (1.5).

1.1 What is life?

The quest to understand how life formed and how it is distributed in the cosmos naturally relies on the assumption that we know exactly what it is! In fact, it is surprisingly difficult to devise a unique, precise definition that includes all that we consider to be living and excludes all that we consider to be non-living. One may attempt to define life (as we know it) in terms of its observed empirical properties, such as its reliance on organic chemistry, its highly-ordered molecular structure, and its ability to metabolize, reproduce and evolve. But there are obvious caveats to such definitions if considered individually: organic matter is not necessarily either living or a product of life; crystals are highly ordered but not considered to be alive; a purely metabolic definition could include fire as a living entity but exclude viruses; and non-living structures such as micelles are capable of mimicking biological replication. It may be the case that we simply do not yet have a sufficiently deep understanding of life's properties to compose an ideal definition. An analogy suggested by Cleland and Chyba (2002) considers how one might attempt to define water in an era before the existence of molecular theory: one could again adopt an empirical approach—water is a clear, odorless, tasteless liquid—but this is not necessarily either unique or universal. Modern science provides a simple, robust definition—water is H_2O—that is meaningful beyond the confines of our home planet and our local experience.

For the time being, we are obliged to accept an empirical definition of life that is consistent with the only example available to us. *Life as we know it* is a highly ordered, carbon-based system that utilizes liquid water as the primary medium for chemical reactions, is capable of processes that we consider to be biological in essence, such as metabolism, photosynthesis, replication and Darwinian evolution, and is hosted by a rocky planet in a nearly-circular orbit around a normal star at a distance that allows water to be present in liquid form on its surface. It requires several distinct classes of molecular polymers that enable it to function, including genetic material composed of nucleic acids, proteins and enzymes composed of amino acids, and membranes containing fatty acids, and it uses adenosine triphosphate as a vital source of chemical energy. Finally, life is homochiral: it utilizes molecules that have right and left-handed forms (enantiomers), with the nucleic acids containing only the right-handed form of the appropriate sugar (ribose in RNA, deoxyribose in DNA), and the proteins and enzymes containing left-handed amino acids (all amino acids except the simplest, glycine, are chiral). All known life forms on our planet share these basic common properties, from the smallest microbes to the tallest trees and the smartest bipeds.

Such a lengthy and multifaceted definition inevitably leads to fundamental questions concerning the origin and universality of terrestrial life. Nucleic acids and proteins carry out distinct but complementary roles in modern terrestrial biology, analogous to the roles of software and hardware in a computer: did this synergy arise spontaneously, or did one function give rise to the other? Which process became established first, replication or metabolism? How did homochirality arise and is the handedness of terrestrial life unique? How did the first cells form? Is carbon-based life the only possible form? Is H_2O the only viable solvent? Are earthlike planets the only suitable hosts? And is it appropriate, in our search for extraterrestrial life, to assume that it has the same fundamental properties as our own?

Several of these questions are at the focus of current research and will be considered further in the following chapters. However, the question of whether life must be carbon-based should be addressed immediately as it affects much that follows. The most commonly discussed alternative to carbon is silicon.

1.2 Could life be silicon-based?

Like carbon (C), silicon (Si) is tetravalent, i.e. it has four electrons available to form covalent chemical bonds, and so any organic molecule might, in principle, have an equivalent form in which Si substitutes for C: for example, silane (SiH_4) is the silicon-based equivalent of methane (CH_4). Also of interest are molecules that contain both Si and C (organosilicon compounds), which can be synthesized in the laboratory and have industrial applications. However, a crucial problem with Si as a basis for life is that none of these compounds typically occur in nature, and even those that can be created synthetically have very limited variety and functionality in comparison to carbon-based organic chemistry. An important reason for this is the weakness of the Si–Si bond, a factor of about two weaker than the C–C bond. Thus, Si does not bond readily with itself to form stable silicon-based equivalents of many biologically relevant moieties, such as the chains and rings found in aliphatic and aromatic hydrocarbons. Moreover, Si cannot form double or triple bonds either with itself or with other elements, which excludes the possibility of Si-based equivalents of biologically important units such as the carbonyl $(C = O)$ and nitrile $(C \equiv N)$ groups. However, Si does bond readily with O to form quartz, silicates and other oxygen-rich mineral compounds, typically arranged in crystalline forms rather than in structures that might have biological functionality. Si is plentiful on Earth—indeed, it is about 200 times more abundant than C in the Earth's crust—but it is found almost exclusively in rock-forming minerals. As Trimble (1997) has remarked, 'You know what a large assemblage of molecules based on silicon looks like—it is piled over every beach.'

Looking beyond the Earth, available evidence suggests that silicon's affinity for oxygen is ubiquitous. Physical samples from within our solar system (such as meteorites, cometary dust, and lunar rocks returned by the Apollo and Luna missions) contain Si in compounds that do not differ radically from those indigenous to Earth. Spectroscopy enables Si-bearing compounds far beyond our solar system to be characterized as well, including those in the interstellar clouds from which new planetary systems are being born, and they are found to be composed almost entirely of silicates such as enstatite $(MgSiO_3)$ and fosterite (Mg_2SiO_4) that form a major component of interstellar dust. An exception occurs in the atmospheres of certain 'red giant' stars enriched in carbon, where observations show silicon carbide (SiC) to be the preferred form. More generally, environments favorable to the production of a limited number of organosilicon compounds might be possible, but this would require a highly unusual local distribution of the elemental abundances, with a dearth of free oxygen. Given the ubiquity of O, which is third on the list of the most abundant chemical elements in the cosmos (see chapter 2), such environments seem likely to be extremely rare. Similar objections apply to other tetravalent elements such as germanium (Ge), which has the additional disadvantage of being very much less abundant than either C or Si. As Carl Sagan remarked many years ago in his book *The Cosmic Connection* (1973), in attempting to shed geocentric 'chauvinisms' when considering the possibilities for extraterrestrial life, carbon chauvinism is the one that seems hardest to overcome. In summary, all the available evidence suggests that carbon-based organic chemistry is the only realistic foundation for naturally occurring biological life in the Universe.

Another possibility that will not be discussed further in this book is that living entities might arise non-biologically from artificial intelligence, i.e. from sophisticated electronic networks that typically utilize the elements Si and Ge in forms that would not occur in nature. Doomsday scenarios that suggest a potential threat to the future of human life from this source have gained media attention. A related question is the possibility of pervasive, self-replicating spacecraft (von Neumann probes), the apparent absence of which is sometimes put forward as evidence against the existence of extraterrestrial intelligence (section 9.3.3). See Hibbard (2014) and Barlow (2013) for further reading on these topics.

1.3 Time constraints

When did life on Earth first arise? Figure 1.1 provides a schematic overview of what is known about the timing of major events in the Earth's geological and biological histories, from its formation some 4.5 billion years ago to the present day. In this section we focus on the first billion years of this timeline, covering the Hadean and early Archean geologic eons, in an attempt to pinpoint the time at which the first life appeared.

Several lines of evidence allow this question to be addressed. Firstly, the age of the Earth itself is reasonably well constrained at about 4.56 ± 0.05 Ga. This result is based on radiometric dating, the standard technique used to determine the ages of rocks and other materials in which trace radioactive impurities were incorporated when they formed: see Dalrymple (2001) for detailed discussion. The abundance of a selected radioactive isotope within the material is compared to the abundance of its decay product, which forms at a known constant rate of decay, and this enables the age to be calculated. The oldest materials available for study turn out to be a class of meteorites (carbonaceous chondrites) considered to be remnants of the planet-forming process in our Sun's protoplanetary disk (see chapter 4). The Sun and planets are thought to have formed over a (geologically) short timescale, within the range of uncertainty quoted above, and hence the age of these meteorites is taken as a reliable estimate of the age of the Earth and other planets in our solar system. The result is consistent with the oldest known indigenous terrestrial minerals (zircon crystals) and lunar rock samples, which have radiometric ages of about 4.4 Ga. The

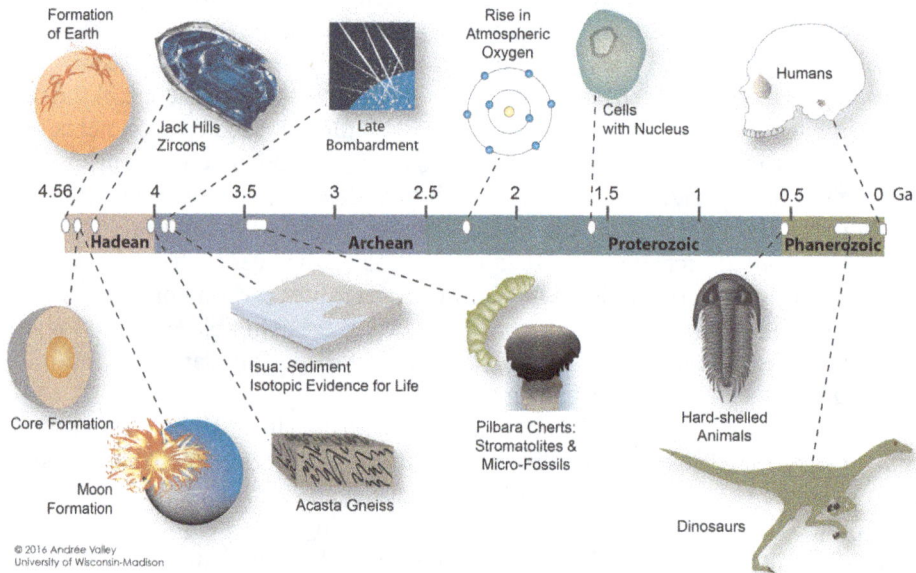

Figure 1.1. Schematic timeline of Earth's history from formation to the present day. Major geologic time periods (the Hadean, Archean, Proterozoic and Phanerozoic eons) are indicated, together with the timings of major events and the formation times of important samples, such as ancient zircons (section 5.3.1) and stromatolite fossils (section 6.2.2). Image credit: Andrée Valley, University of Wisconsin-Madison.

absence of surviving rocks in the Earth's crust that date all the way back to its epoch of formation (or, more realistically, to the Moon-forming event—see below) is not surprising, given that it is being renewed by geological activity on timescales short compared with its age.

How soon after the Earth formed did conditions become suitable for life? And once the environment was suitable, how long did it take for life to actually emerge? The quest to answer these questions is a major area of current research in astrobiology. To address them we must attempt to gather evidence relating to not only the Earth itself but also its planetary environment. The Earth is unusual in having a relatively large moon (27% by radius relative to the planet, compared with <5% for those of the other major planets in our solar system). Our Moon is thought to have formed some 4.5 Ga ago, the result of a giant impact between the Earth and a Mars-sized protoplanet. This catastrophic event effectively resurfaced the Earth, creating an 'ocean' of magma, and thereby erasing any record of conditions that prevailed at earlier times. The time required to reach habitable conditions after the Moon-forming event is uncertain, but recent research suggests that it may have been reasonably short: geochemical studies of ancient minerals indicate that surface water and solid continents were present as early as 4.2–4.4 Ga ago, suggesting reasonably clement conditions. See chapter 5 for further discussion.

Another important consideration is the enduring threat of further disruption by cosmic impacts. Geologic activity and weathering have erased the record of ancient impact events on Earth, but our nearest neighbor, the Moon, has preserved a record of past cratering that is likely to be quite similar to that suffered by the Earth. A careful study of the cratering record of the Moon, in combination with geochemical and radiometric analyses of samples returned by space missions, indicates that the most intense cratering occurred prior to the formation of the basaltic plains (dark areas) on the Moon some 3.2–3.8 Ga ago. The plains that form the familiar 'face' of the lunar disk were themselves the result of giant impacts, creating basins that subsequently flooded with magma. If impactors similar in size to those responsible for the lunar basins also struck the Earth, it seems probable that any life originating before that time would have been eradicated. Research on the chronology of these basin-forming impacts suggests that several may have arrived over a relatively short time period, some 3.8–3.9 Ga ago, perhaps as the result of a disturbance in the outer solar system that led to an epoch of cratering termed the *late heavy bombardment*. We return to this topic in section 5.2.

Evidence for the earliest extant life is sought by investigating structural and chemical signatures of life embedded in rocks from the oldest surviving areas of continental crust. Microfossils (fossilized microbes and microbial colonies such as stromatolites) dated in the range 3.5–3.7 Ga have been identified in samples from Australia and Greenland (e.g. Nutman *et al* 2016). Additional evidence is provided by data on carbon isotope abundances: because of its slightly greater mass, carbon-13 (^{13}C) tends to be less favored in biological reactions relative to the more common isotope (^{12}C), and hence organic matter formed biologically tends to favor ^{12}C over ^{13}C, a process referred to as fractionation. Tentative evidence for this and other signatures of biological activity has been reported in sediments dated at 3.85 Ga.

To summarize, the epoch for the origin of life is conservatively constrained to the range 3.5–4.4 Ga; however, a case can be made for a much tighter limit of 3.85 ± 0.10 Ga, constrained by probable fossil and isotopic evidence on one side, and cessation of potentially planet-sterilizing impacts on the other. In any case, life appears to have arisen reasonably quickly once conditions stabilized, on timescales less than (and perhaps *much* less than) 20% of the age of the planet.

1.4 Scenarios for the origin of life on Earth

What evolutionary processes led from the abiotic early Earth to the emergence of microbial life? It is helpful to consider this question in three simple stages, outlined below and discussed in greater detail in chapter 6:

1. The prebiotic world (production and accumulation of organic building blocks).
2. The protobiotic world (polymerization of RNA, peptides; first biochemical cycles).
3. An organizational phase in which the first living cells appeared.

1.4.1 The prebiotic world

The prebiotic world takes as its starting point an epoch where surface water is present, together with an atmosphere containing gases such as CO, CO_2, CH_4, N_2 and NH_3. The inventory of potentially life-relevant molecules is steadily enriched, partly by delivery of preformed organic species carried by interplanetary debris falling to Earth, and partly by chemical reactions occurring *in situ* in the atmosphere, in shallow surface water, in crustal fissures or cavities, or in hydrothermal systems in the deep ocean. Endothermic reactions leading to greater molecular complexity may be driven by energy sources, as available, in each of these environments, which might include solar radiation, geothermal heat, chemical energy, kinetic energy released by cosmic impacts, and lightning discharge. The products of this phase are presumed to include amino acids, lipids, sugars, nucleic acid bases, and, ultimately, the first nucleotides.

One of the first practical demonstrations of abiotic synthesis of biologically relevant molecules was the well-known Miller–Urey experiment (figure 1.2), carried out in its original form in 1952. An atmosphere of H_2, H_2O, NH_3 and CH_4 was subject to an electrical discharge that simulated lightning. The ensuing chemical reactions produced an array of products, ranging from HCN and H_2CO to amino acids and fatty acids. The experiment thus demonstrated that, under the assumed conditions, the production of at least some of the molecules required for life is facile. However, the composition of the initial atmosphere has been called into question by more recent research. Any primary atmosphere existing prior to the Moon-forming impact would have been lost and subsequently replaced by a secondary atmosphere that accumulated from volcanic outgasing. As the Earth's gravity is too weak to retain free hydrogen over significant timescales, it would be abundant only if it were continuously replenished. Available evidence indicates that the secondary atmosphere was probably quite hydrogen-poor, not only lacking H_2 but also carrying the bulk of its carbon and nitrogen in the form of CO_2 and N_2, respectively, rather than

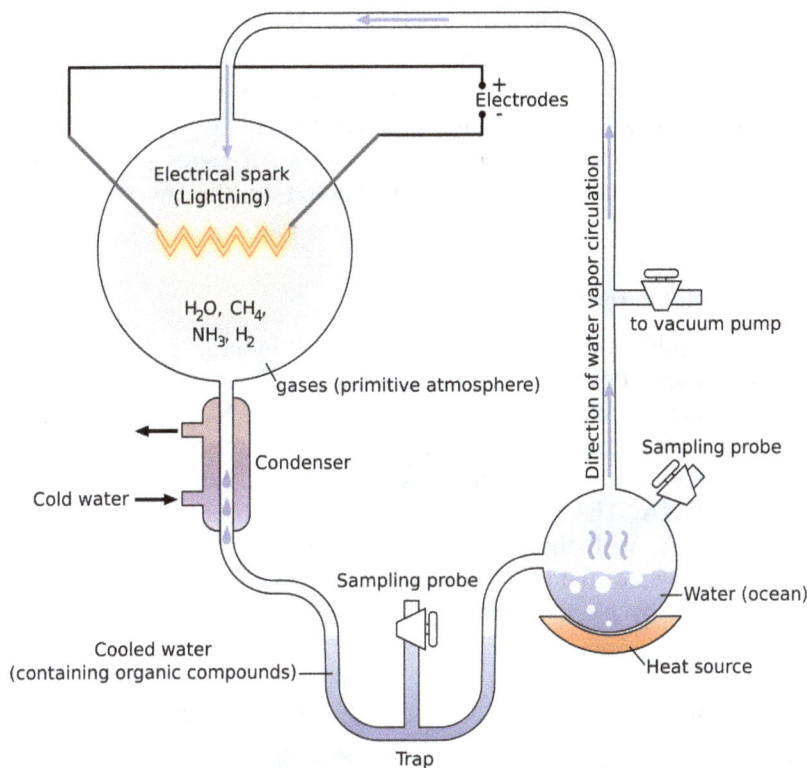

Figure 1.2. Schematic illustration of the Miller–Urey experiment, in which amino acids and other organic compounds are synthesized abiotically when an atmosphere of H_2, H_2O, NH_3 and CH_4 is subject to an electrical discharge that simulates lightning. Image credit: YassineMrabet, Wikimedia Commons.

CH_4 and NH_3 (see section 5.3). The initial step in the process of abiotic synthesis in the original Miller–Urey experiment is the production of radicals from the starting molecules, for example:

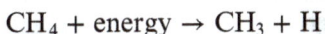

$$CH_4 + energy \rightarrow CH_3 + H$$

and similarly NH_2 and OH. Being reactive, these radicals then combine with each other to form products of higher molecular weight. But when the experiment was repeated with CO_2 and N_2 as the initial bearers of C and N instead of CH_4 and NH_3, the prebiotic yield fell dramatically, by orders of magnitude, because CO_2 and N_2 are much more tightly bound, less reactive molecules. Thus, in summary, the Miller–Urey experiment is a helpful demonstration of principle, but probably not an accurate simulation of prebiotic synthesis on the early Earth.

The Earth is, by cosmic standards, a remarkably hydrogen-poor environment (see chapter 2), yet one lesson that the Miller–Urey experiment has taught us is that an abundance of free hydrogen and/or hydrogenated molecules greatly facilitates prebiotic synthesis. Two natural solutions to this apparent dilemma can be considered (section 5.4): exogenous delivery of organic matter from space, or efficient endogenous processes that utilize the most plentiful source of hydrogen on Earth: H$_2$O. A demonstration of principle for the delivery route arises from the detection of extraterrestrial amino acids and other prebiotic molecules in meteorites. Promising scenarios for *in situ* production include the deep-ocean hydrothermal systems, in which organic compounds are synthesized at the dynamic interface between hot hydrothermal fluids and cool seawater. It seems probable that both exogenous and endogenous processes contributed to the inventory of organic molecules on the early Earth.

1.4.2 The protobiotic world

The products of prebiotic chemical evolution may undergo polymerization under favorable conditions. Polymerized products vital to biology include nucleic acids (figure 1.3) and amino acid chains (peptides; figure 1.4). Nucleic acids are made up of a 'backbone' of alternating sugar and phosphate molecules, with nitrogenous nucleobase molecules bonded to each sugar, the sugar being ribose in RNA and deoxyribose in DNA. The bases are guanine (G), cytosine (C), adenine (A) and uracil (U) in RNA, with thymine (T) substituting for uracil in DNA. DNA is structured in the form of the well-known double helix (figure 1.3), in which the two strands are joined by electrostatic (hydrogen) bonds between paired bases: the

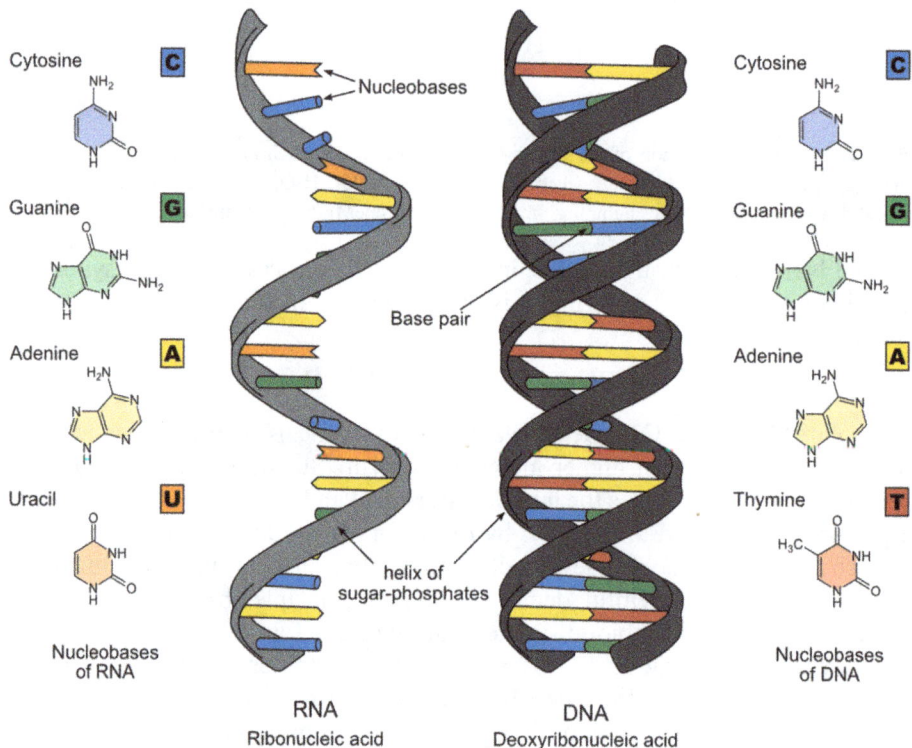

Figure 1.3. Comparison of the structure of RNA and DNA. The helical backbone is composed of alternating sugar and phosphate units, with bases attached to each sugar. The base pairing rules enable replication and transcription. Image credit: Sponk, Wikimedia Commons.

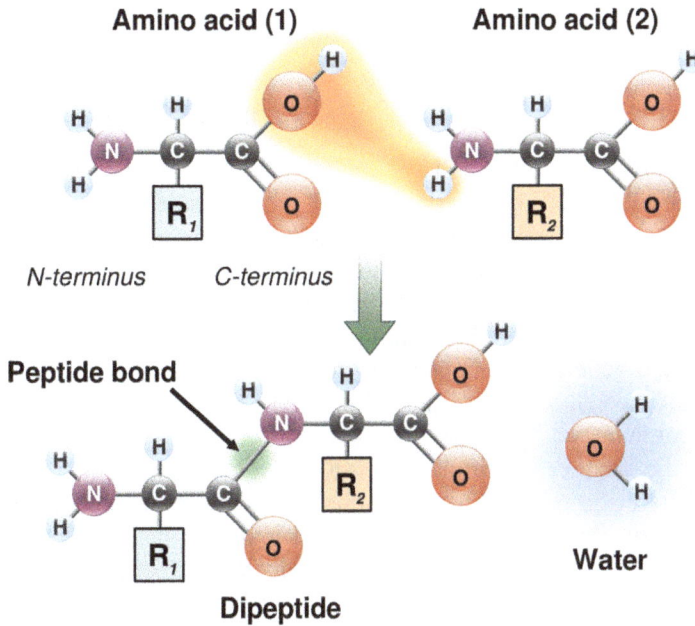

Figure 1.4. Illustration of the molecular structure of amino acids and the formation of a dipeptide from two amino acids. Each amino acid contains an amine (NH_2) and a carboxyl (COOH) group joined by a carbon atom. R_1 and R_1 represent different side chains: in glycine, for example, the side chain is H, and in alanine it is CH_3. Image credit: YassineMrabet, Wikimedia Commons.

geometrical distribution of hydrogen bond 'donors' and 'acceptors' in each base determines paring rules, such that A pairs with T, and C pairs with G. Unlike DNA, RNA does not generally link with another strand to form a double helix. DNA replicates by division of the two strands, followed by attachment of additional nucleotides, each strand forming a copy dictated by the base-pairing rules. DNA also transcribes RNA (with A paired to U instead of T). RNA regulates the formation of peptides from individual amino acid monomers (figure 1.4), and peptides are the building blocks of proteins and enzymes.

As noted earlier in this chapter, nucleic acids and proteins carry out distinct but vital roles in modern biology, including storage of genetic information (DNA) and catalysis of metabolic reactions (proteins), and this presents a problem in understanding whether one function preceded the other or whether both could have arisen simultaneously. A possible solution to this question that has been at the core of origins of life research since publication of a seminal paper by Gilbert (1986) is the RNA world hypothesis. Based on the versatility of RNA, a polymer that has an inherent capacity for information storage as well as the ability to act as a catalyst, a primitive precursor to modern biology is proposed in which RNA carries out both of these functions. The hypothesis is not universally accepted, however; alternative suggestions include models that propose a prominent early role for peptides and primitive metabolic cycles that predate the existence of genetic material. See section 6.1 for further discussion.

1.4.3 Organization

The steps discussed above suggest routes to the production and polymerization of biologically-relevant molecules by abiotic processes occurring in some prevalent planetary environment that happens to be favorable. However, life on Earth is composed of *cells*—in effect, small containers, each enclosing a microenvironment finely tuned for biological functionality. A cell is the smallest autonomous unit of a living system that can replicate independently by division. The first life was composed of the simplest (prokaryotic) cells, typically about a micrometer in size, in which a fluid (cytoplasm) is contained within a cell wall. The fluid contains genetic material (a primitive 'nucleoid' containing a single chromosome of DNA) and ribosomes, which generate proteins from RNA templates. The cell wall maintains the structure of the cell and serves as a filter, separating the internal fluid from the external environment.

How did the first cells form? Cell walls contain membranes composed of phospholipids (section 6.1.4). Each phospholipid contains a phosphate 'head' and two fatty acid 'tails', which are hydrophilic and hydrophobic, respectively. Because of this, phospholipids naturally self-organize in the presence of water to form structures such as micelles and liposomes, with the heads pointing outward and the tails pointing inward. A liposome is a small bubble bound by a spherical lipid bilayer, resembling a simple cell wall, and is thus a natural candidate for the structural unit of the first cells. Liposomes may readily form in turbulent water, given a sufficient concentration of phospholipids, but whether this was the case on the early Earth is an open question. Phosphate minerals are present in the Earth's crust, and some fatty acids would have been available as the result of endogenous production and/or exogenous delivery (they are detected in trace amounts in meteorites); but they may not have accumulated in high enough concentrations for efficient phospholipid production. An alternative proposal is that micrometer-sized cavities in common porous minerals such as feldspars may have served as an initial means of compartmentalization, prior to the development of phospholipid membranes.

1.4.4 Life from elsewhere?

An alternative view on the origin of terrestrial life is the panspermia hypothesis, which proposes that life formed elsewhere in the cosmos and was subsequently delivered to the Earth by some mechanism. It is notable that this hypothesis does not address the question of how the first life arose; instead, it attributes it to some other place and time. The most widely discussed version of the hypothesis is a 'random accretion' panspermia, in which dormant microbial life is transported by cosmic debris such as meteors or cometary dust, the arrival of which seeded the Earth with its first life[1]. The exchange of materials between planetary bodies within our solar system is confirmed by the discovery of meteorites on Earth shown to be

---

[1] Other versions include proposals that the Earth might have been seeded (intentionally or accidentally) by some form of extraterrestrial intelligence.

impact-ejecta from the Moon and from Mars: i.e. crustal material from those bodies that was launched into space by cosmic impacts, some fragments of which eventually fell to Earth. However, no credible evidence has been presented for the presence of extraterrestrial life, either extant or extinct, in any meteorite or interplanetary dust sample studied to date, regardless of its origin (see section 7.2.2 for further discussion of Martian meteorites).

Nevertheless, the possibility of panspermia as the source of life on early Earth is not ruled out. Current experimental research focuses on the ability of extremophiles to survive in space environments and during planetary impact events. The results are of general interest, not only as a test of the panspermia hypothesis but also to provide guidance on planetary protection for the space program—to ensure, for example, that our species does not inadvertently contaminate Mars with terrestrial life prior to a comprehensive search for endogenous Martian life. The primary hazard to organisms in space is ultraviolet radiation, which tends to destroy them on timescales far shorter than typical timescales for interplanetary transport, unless they are very well shielded within the carrier material. Simulated planetary impact experiments suggest that organisms may survive the final phase of the delivery process.

1.5 Searching for life beyond Earth

A realistic search for life elsewhere in the Universe requires assumptions to be made on its fundamental properties: what it is made of, how it functions, and what conditions are needed for it to form and survive. Because we have only one instance of life to use as a template, we have little choice but to assume that its most basic properties are not radically different from those of terrestrial life (section 1.1), i.e. a highly-ordered system based on organic chemistry, optimized for conditions that allow water to exist in liquid form, and capable of biological functions such as metabolism and photosynthesis that generate measurable products. These assumptions allow specific strategies to be developed to search for life, both within and beyond our solar system.

A prime reason why the Earth is a suitable host for life is its location in the 'Goldilocks zone' of our solar system, i.e. in a nearly-circular orbit of appropriate radius around a stable, middle-aged star, where it is neither too hot nor too cold for the bulk of its water to be in liquid form (see section 4.4). By contrast, our nearest planetary neighbors are far less well placed. Venus, similar to Earth in terms of size and mass and only 30% closer to the Sun, has suffered a runaway greenhouse effect leading to surface temperatures in excess of those tolerated by even the hardiest hyperthermophiles. Mars, 50% further from the Sun than Earth and only about a tenth of its mass, has suffered the opposite fate, failing to sustain an atmosphere dense enough to provide a greenhouse effect that might have enabled it to stay warm and wet. As we explore exoplanetary systems in search of planets that might be good candidates to host life, we naturally focus on potential analogs of the Earth—planets similar in terms of location (in the Goldilocks zone of the system) and mass (possessing sufficient gravity to retain a reasonably dense atmosphere).

A useful approach to our investigation is to consider possible answers to the following question: how would alien astronomers observing the Earth from some remote location be able to determine whether it hosts life? One possibility might be for them to tune in to your favorite radio station, if they possess a sufficiently sensitive receiver, but let us turn the clock back a 100 years to a time before this technology existed on Earth. Let us also assume that the aliens' telescopes cannot resolve specific details of the surface, but are equipped with spectrometers capable of detecting atmospheric gases on a global scale. The observations would reveal that the Earth's atmosphere contains substantial quantities of molecular oxygen and only trace amounts of CO_2, in contrast to the atmospheres of our nearest planetary neighbors where this situation is reversed. To the best of our knowledge, the only feasible method to produce atmospheric oxygen in such large quantities is photosynthesis, a biological process. Hence, observers familiar with this process would be justified in concluding that they have presumptive evidence for life.

The example described above considers just one of many possible biosignatures that might enable us to detect life beyond Earth. The term 'biosignature' is a broad one that encompasses experimental analyses of samples from candidate hosts as well as remote sensing by spectroscopy. Investigations of potential Martian biosignatures, for example, range from *in situ* experiments carried out by landed spacecraft to laboratory studies of Martian materials delivered to Earth (currently limited to Martian meteorites, with plans for future targeted sample return missions), as well as remote spectroscopic observations of the surface and atmosphere carried out by both Earth-based and space telescopes. These strategies and techniques are discussed further in the concluding chapters of this book, focusing in turn on Mars, icy moons, and exoplanetary systems (chapters 7, 8 and 9, respectively).

Questions and discussion topics

- In the mid-1960s the eminent palaeontologist George Gaylord Simpson criticised exobiology as a science 'that has yet to demonstrate that its subject matter exists!'. Consider what arguments you might use to justify modern research in astrobiology if it were subjected to such a criticism.
- Explain the difference between the 'top down' and the 'bottom up' approaches to research on the origin of life, giving examples of each.
- Consider ways in which progress in one field of science relevant to astrobiology might lead to progress in another.
- Assess the realism of the Miller–Urey experiment as a demonstration of how important biomolecules came to exist on the early Earth.
- Solar ultraviolet (UV) radiation is one possible source of energy to drive chemical reactions on early Earth. Do you expect that the average intensity of solar UV reaching the surface of the Earth during the first 500 million years of its history would have been greater, smaller, or very similar to that today? On what does the answer depend?
- Identification of Martian meteorites on the surface of the Earth suggests the possibility that life on Earth actually originated on Mars and was carried here

through interplanetary space. Do you think it is equally probable that future exploration of the Martian surface will reveal meteorites originating on Earth that might have carried life in the opposite direction?

References and further reading

Bergin E A 2014 Astrobiology: An astronomer's perspective *AIP Conf. Ser.* **1638** 5

Barlow M T 2013 Galactic exploration by directed self-replicating probes, and its implications for the Fermi paradox *Int. J. Astrobiol.* **12** 63

Chyba C F and Hand K P 2005 Astrobiology: The study of the living universe *Ann. Rev. Astron. Astrophys.* **43** 31

Cleland C E and Chyba C F 2002 Defining life *Orig. Life Evol. Biosph.* **32** 387

Dalrymple G B 2001 The age of the Earth in the twentieth century: A problem (mostly) solved *Geol. Soc. London Spec. Publ.* **190** 205

Domagal-Goldman S *et al* 2016 The astrobiology primer v2.0 *Astrobiology* **16** 561

Gilbert W 1986 Origin of life: The RNA world *Nature* **319** 618

Hibbard B 2014 Ethical artificial intelligence http://arxiv.org/abs/1411.1373

Nutman A P, Bennett V C, Friend C R L, van Kranendonk M and Chivas A R 2016 Rapid emergence of life shown by discovery of 3,700-million-year-old microbial structures *Nature* **537** 535

Sagan C 1973 The Cosmic Connection: An Extraterrestrial Perspective (Cambridge: Cambridge University Press)

Trimble V 1997 Origin of the biologically important elements *Orig. Life Evol. Biosph.* **27** 3

Chapter 2

Cosmic synthesis of the biologically important chemical elements

Astronomy has shown that the observable Universe is composed predominantly of hydrogen and helium[1]. Indeed, according to modern cosmological theory, when it was born some 14 billion years ago matter in the Universe was to all intents and purposes *just* hydrogen and helium. Enrichment with the heavier elements needed to make rocky planets, organic molecules and carbon-based life occurred gradually over time, as they were synthesized inside stars and injected into interstellar space. Even so, the combined mass of all elements heavier than H and He amounts to only about 2% of the total after 14 billion years of evolution. Which of these elements are essential to life? How were they formed and distributed, so that they became available for inclusion in subsequent generations of stars and their planetary systems? How long did it take for them to become abundant enough for life to be possible? This chapter reviews the extent to which we are able to answer these questions.

2.1 Elements essential to life

A shortlist of the chemical elements essential to life as we know it appears in table 2.1. Inclusion is based on the vital structural and functional roles these elements play in biomolecules and biochemical processes: a more extensive list might include additional trace elements that have lesser or more limited biological roles, such as certain nutrients typically found in a daily multivitamin and mineral supplement (see, for example, Trimble 1997). The relative abundances of these

---

[1] Only baryonic matter, i.e. matter composed of the known chemical elements, is considered here. Exotic forms invoked to explain 'dark matter' are beyond the scope of this book.

doi:10.1088/978-1-6817-4676-0ch2

Table 2.1. A shortlist of chemical elements essential to life.

Element	Z	Usage
H	1	Water, organic chemistry
C	6	Organic chemistry
N	7	Organic chemistry
O	8	Water, organic chemistry; aerobic processes
Na	11	Salt; fluid regulator
Mg	12	Chlorophyll
Si	14	Cell walls, bones, diatoms
P	15	Nucleic acids, phospholipids; energy transport
S	16	Proteins, enzymes; anaerobic respiration
Cl	17	Salt; electrolyte; metabolic role
K	19	Fluid regulator; fertilizers
Ca	20	Bones; various biochemical roles
Mn	25	Enzyme cofactors
Fe	26	Hemoglobin; enzyme cofactors
Cu	29	Proteins; aerobic respiration
Zn	30	Enzyme cofactors; various biochemical roles
Mo	42	Enzyme cofactors
I	53	Thyroid hormones

and other chemical elements in our solar system[2] are illustrated in figure 2.1, displayed in order of increasing atomic number (Z). It is notable that the 'CHON' elements utilized by organic chemistry make up four of the six most abundant elements, the other two being the (chemically inert) noble gases He and Ne. For simplicity, only the first 35 elements are shown in figure 2.1: with the exceptions of Mo and I, those of higher Z appear to be of little or no direct relevance to life.

Figure 2.1 provides clues as to how the elements formed and evolved. The overall trend is a general decline from abundant light elements to rare heavy elements, with a range that spans many orders of magnitude. There are evident departures from the general trend, however; these include (i) a dramatic trough centered at $Z = 4$ (Be), (ii) a pattern for $Z \geqslant 6$ of even-numbered elements being more abundant than adjacent odd-numbered ones, and (iii) a broad peak centered at $Z = 26$ (Fe). As discussed in the following section, these departures reflect the relative stability of the atomic nuclei and the nature of the mechanisms that produce them: in particular, elements formed by combining α particles (^{4}He nuclei), corresponding to even numbered increments in Z, are generally more stable and more readily produced than the intermediate cases.

---

[2] The mean abundances in our solar-system are naturally dominated by the Sun, deduced from observations of the solar atmosphere, and augmented by analyses of meteorites and the Earth's crust for many of the heavier elements, all being normalized relative to silicon $= 10^6$. The results are thought to reflect the elemental composition of the interstellar cloud that gave birth to the solar system.

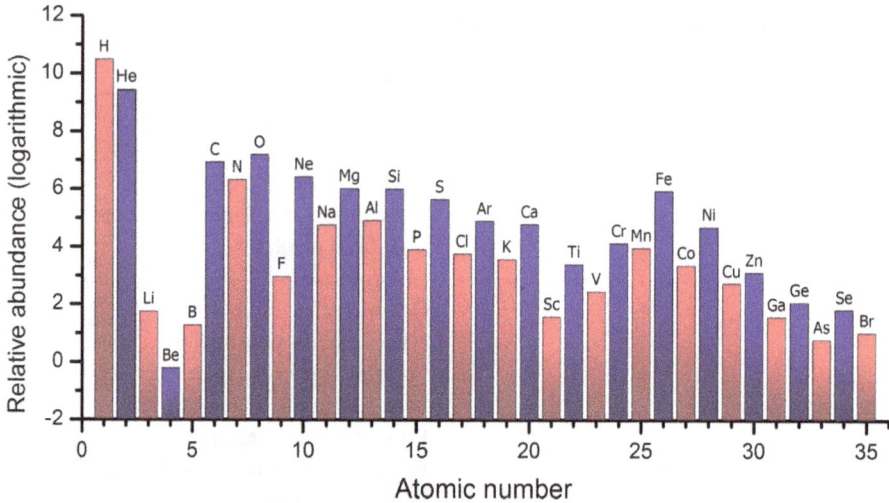

Figure 2.1. Histogram displaying mean relative abundances of chemical elements in the solar system. The first 35 elements are shown in order of increasing atomic number. Abundances are logarithmic by number of atoms relative to $Si = 10^6$, based on data from Asplund *et al* (2009). Columns for odd and even-numbered elements are colored pink and blue, respectively.

2.2 Origin of the elements

Simultaneous advances in cosmology, nuclear physics and stellar astrophysics during the 20th century have enabled robust models to be developed to account for the origins of the chemical elements and their relative abundances. According to the standard cosmological model, the Universe was initially composed of H and He, and the heavier elements were generated later, primarily by fusion reactions (nucleosynthesis) inside stars. Figure 2.2 shows the periodic table of the elements with color coding to indicate the principal source(s) of each element. This section reviews the main processes responsible for this pattern, and the typical timescales on which they operate, with particular attention to the elements identified in table 2.1 as essential to life.

2.2.1 A primer on nuclear fusion

A detailed discussion of the relevant nuclear physics is beyond the scope of this book, but a few basic principles may be helpful. The chemical properties of an element are determined by its atomic number, which is equal to the number of protons (p) in the nucleus. Many elements have stable isotopes that differ from each other only in atomic mass, because their nuclei contain different numbers of neutrons (n). For example, hydrogen, which in its common form has a nucleus containing just a single proton, also has heavier isotopes deuterium (p + n) and tritium (p + 2n). Nuclear reactions that produce these and other isotopes are essential initial steps along the way to fusion of heavier elements[3].

---

[3] As discussed elsewhere in this book, isotopes also provide helpful diagnostic tools for studying many processes relevant to astrobiology, because the mass differences result in fractionation (measurable differences in isotopic abundance ratios) if the rate of a reaction depends on mass.

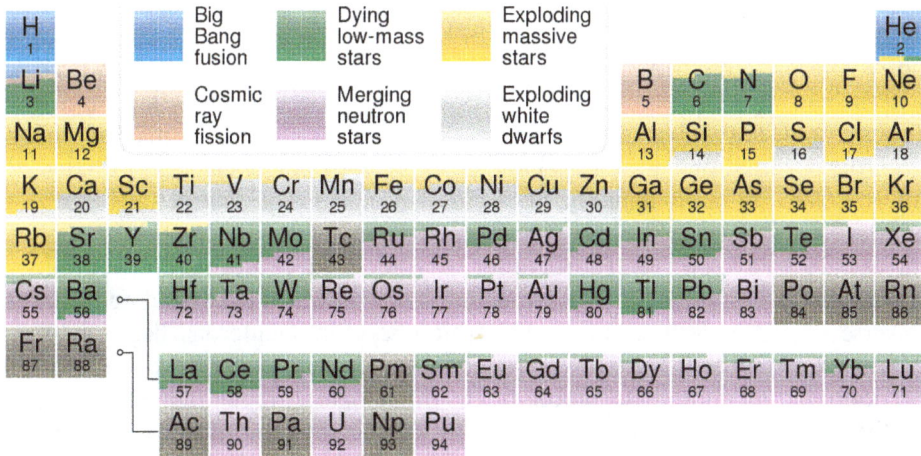

Figure 2.2. Periodic table of the elements, indicating their primary sources. The cross-over between low-mass and massive stars occurs at about 5 solar masses. Image credit: Cmglee (Wikimedia Commons), based on data by Jennifer Johnson at Ohio State University.

The nuclei of heavier elements and isotopes are synthesized inside stars by sequences of reactions that fuse those of lighter elements together. These reactions are generally exothermic, i.e. there is a net release of nuclear binding energy. This energy is the source not only of a star's heat and luminosity but also of outward thermodynamic pressure that maintains its stability by balancing the inward force of gravity. Consider, for example, fusion of two protons to form a deuteron (deuterium nucleus, d) and a positron (e^+):

$$p + p \rightarrow d + e^+ + \text{energy.} \tag{2.1}$$

Although the reaction is exothermic, it is difficult to activate because protons bear positive charge: there is an electrostatic repulsion between them arising from Coulomb's law, and this tends to inhibit the reaction. An initial input of energy is needed to push them close enough together for the short-ranged nuclear forces to overcome electrostatic forces: the situation is analogous to a golfer putting a ball into a hole at the top of a slope. Because of this, nuclear fusion reactors need to maintain very high temperatures and pressures. However, an important exception arises for reactions involving neutron capture, as they have no Coulomb barrier (because the neutron bears no charge), for example:

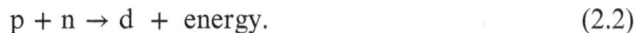

$$p + n \rightarrow d + \text{energy.} \tag{2.2}$$

As discussed below, p + n fusion was crucial in the very early stages of the Universe, whereas p + p fusion is important inside stars. Free neutrons, i.e. those not contained within a nucleus, are unstable (decaying to a proton and an electron on a timescale of about 15 min), and are therefore unavailable except in situations where they are being rapidly produced.

2.2.2 Element production in the early Universe

The evolutionary (big bang) model for the birth of the Universe is generally accepted as it successfully predicts many of its observed properties, including large scale structure and the universal expansion (Hubble's law), the cosmic microwave background radiation, and the abundances of the light elements. Initially, the Universe was an unimaginably hot and dense soup of interacting subatomic particles. As it expanded from the original singularity, the temperature of the Universe fell below $\sim 10^{10}$ K, sufficient for baryonic matter (protons and neutrons) to become stable. For a few minutes thereafter, as it continued to expand and cool, the Universe was effectively a primordial fusion reactor. Deuterons began to form, principally by proton–neutron capture (equation (2.2)); initially the deuterons were also destroyed by photodissociation, but as the temperature fell they became stable and their numbers began to rise. Deuterons themselves then underwent neutron or proton capture to form ^{3}H or ^{3}He nuclei, and a further capture event led to ^{4}He, a highly stable product. The primordial abundance of helium relative to hydrogen was fixed by the relative numbers of neutrons and protons at the commencement of nucleosynthesis (approximately 1/6), as free neutrons are captured and processed to ^{4}He on a timescale short compared with their timescale for decay.

H is the only element of direct relevance to life that was produced in the big bang. Apart from H, He, and their isotopes, the only other elements formed were ^{7}Li and ^{7}Be in tiny amounts (abundances $\sim 10^{-10}$ relative to H). This situation arose because there are no stable elements of atomic mass 5 or 8: ^{4}He nuclei (α particles) cannot form stable higher-mass products by single-capture events involving protons, neutrons, or other ^{4}He nuclei. The production of carbon, in particular, requires the quasi-simultaneous fusion of *three* ^{4}He nuclei (the triple-α process, see below), for which the necessary temperatures and densities must be sustained over time periods very much longer than those available in the early Universe. The rates of all fusion reactions drop precipitously to negligible levels as the density of the expanding Universe falls, and after ~ 10 min no further synthesis is likely to occur until the first generations of stars are born.

2.2.3 Element production in stars

The big bang had only a few minutes in which to create elements by fusion, but the lifetime of a star ranges from tens of millions to tens of billions of years (depending on its mass). The evolutionary status of a star is indicated by its locus on the Hertzsprung–Russell (HR) diagram, shown schematically in figure 2.3, which plots mean surface temperature versus luminosity. Our Sun is a main-sequence star, lying on a diagonal band occupied by stars that utilize H as their primary fuel for nucleosynthesis, and generate He as the primary product. The position on the main sequence occupied by a star depends on its mass, with stars more massive and less massive than the Sun lying to the upper left and lower right, respectively.

Nuclear fusion reactions occur in a star's dense core, where temperatures are typically in the range 10–100 million Kelvin, very much hotter than at the surface. In the first main-sequence stars to form, the process was initiated by the p + p reaction

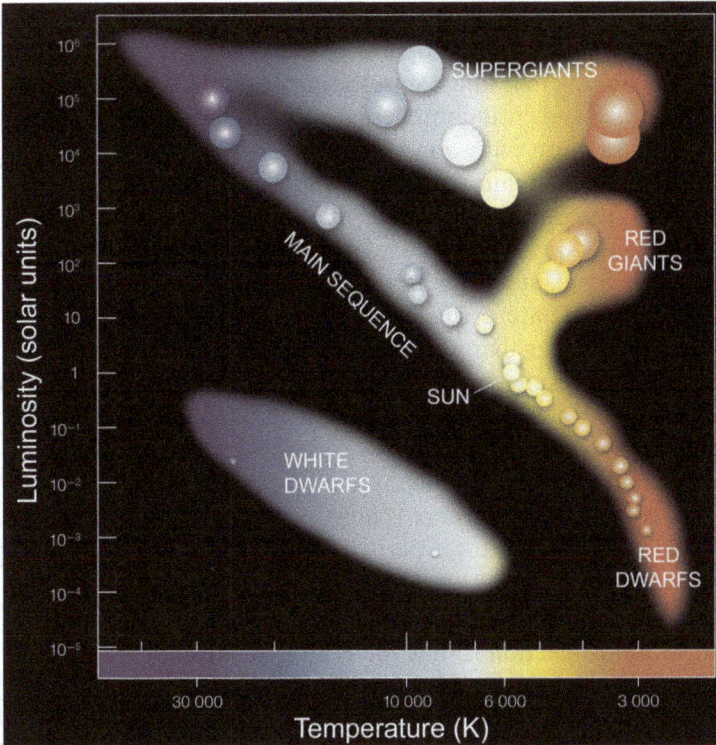

Figure 2.3. A schematic Hertzsprung–Russell (HR) diagram, plotting stellar surface temperature versus luminosity. By convention, temperature is plotted increasing to the left, as it is deduced from an observed quantity (the color index) that increases inversely with the temperature. Image credit: European Southern Observatory, annotated by the author.

(equation (2.1)) to produce d, followed by further proton-capture reactions to produce ^{3}He and then ^{4}He. This sequence (the proton–proton chain) may be summarized:

$$4p \rightarrow {}^{4}\text{He} + 2e^{+} + \text{energy}. \tag{2.3}$$

The proton–proton chain is not the only way to fuse H into He. The main alternative route (the CNO cycle) requires pre-existing ^{12}C to be present to act as a catalyst, which was not the case in the first generations of stars; it became significant later, once ^{12}C had been produced and recycled (section 2.3). There are several steps that may be summarized in two parts:

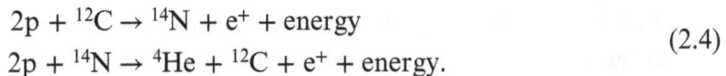

$$2p + {}^{12}\text{C} \rightarrow {}^{14}\text{N} + e^{+} + \text{energy}$$
$$2p + {}^{14}\text{N} \rightarrow {}^{4}\text{He} + {}^{12}\text{C} + e^{+} + \text{energy}. \tag{2.4}$$

The catalytic role of ^{12}C is indicated by the fact that it is both an initial reactant and an end product. The CNO cycle is important from an astrobiological perspective as the primary source of nitrogen. As an intermediary product, ^{14}N is both produced and consumed, but some equilibrium abundance is maintained after initial production.

A star remains on the main sequence until the supply of hydrogen fuel in its core is depleted. The balance between the opposing forces of gravity and thermodynamic pressure is self-regulating in a stable star: the greater its mass, the greater its internal density, temperature, and resultant energy output that counters gravity. For this reason, massive stars 'burn out' sooner, requiring more rapid energy production to maintain equilibrium: the main-sequence lifetimes predicted for those 10–20 times more massive than the Sun (blue stars to the upper left in figure 2.3) are of order 10 *million* years (~10 Ma), compared with 10 *billion* years (~10 Ga) for the Sun itself. On the other hand, stars significantly less massive than the Sun (red dwarfs, lower right in figure 2.3) are predicted to remain on the main-sequence for times comparable with or greater than the present age of the Universe.

The instability that results from the depletion of nuclear fuel causes the core to contract and heat up, balanced by an expansion and cooling of the outer layers, and the star then becomes a red giant or supergiant (upper right in figure 2.3). This transformation continues until an alternative source of energy is ignited to halt the core's collapse. At this stage, the core is mostly composed of helium nuclei, and the logical next step would be for these particles to fuse together to form beryllium:

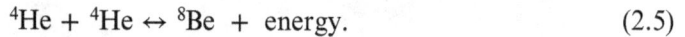

$$^{4}\text{He} + {}^{4}\text{He} \leftrightarrow {}^{8}\text{Be} + \text{energy.} \tag{2.5}$$

The problem with this reaction is that ^{8}Be is a highly unstable isotope (indicated by the two-way arrow), decaying back to helium on a timescale of about 10^{-16} s. But if collisions are frequent enough there is a finite probability that a third ^{4}He nucleus will encounter a ^{8}Be before it decays, allowing ^{12}C to form by what is known as the triple-α process. That this process actually generates non-negligible quantities of C depends not only on the fact that stars can maintain the necessary physical conditions for millions of years, but also on a fortuitous correspondence of energy levels, termed a resonance: the combined energy levels of ^{4}He and ^{8}Be correspond almost exactly to an energy level of ^{12}C, and this allows carbon to be formed in a stable state. The triple-α process is a crucial step in heavy element production, enabling a route not only to carbon itself but also to the heavier elements that depend on the availability of C as a reactant for further fusion[4].

A further capture of an α particle leads to oxygen synthesis:

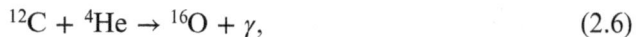

$$^{12}\text{C} + {}^{4}\text{He} \rightarrow {}^{16}\text{O} + \gamma, \tag{2.6}$$

where γ is a gamma-ray photon. However, as helium in the core becomes depleted, the star reaches a new critical point in its evolution, where it must ignite a new fuel to remain stable, or undergo core collapse. Fusion involving C and O as reactants may form elements such as Ne, Mg and Si, but, because of the progressively greater Coulomb barrier that inhibits fusion of nuclei of increasing atomic number, the temperatures and pressures needed to ignite them are correspondingly greater than for H and He. Such conditions cannot be reached in stars below a critical mass of about five solar masses (5 $M_{\odot}$). As energy production shuts down in a less massive

---

[4] The triple-α process may be considered an example of the anthropic principle, which postulates that life could not exist if certain fundamental parameters of the Universe were different.

star such as the Sun, the outer layers are ejected to reveal the collapsed core, which has become a white dwarf, still hot but not luminous (figure 2.3) because of its small size.

Massive stars ignite further cycles of exothermic fusion reactions to produce a variety of heavier elements, several of which are relevant to astrobiology (section 2.1). These reactions include the so-called carbon-burning cycle:

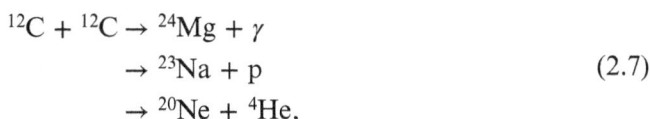

$$
\begin{aligned}
{}^{12}\mathrm{C} + {}^{12}\mathrm{C} &\rightarrow {}^{24}\mathrm{Mg} + \gamma \\
&\rightarrow {}^{23}\mathrm{Na} + \mathrm{p} \\
&\rightarrow {}^{20}\mathrm{Ne} + {}^{4}\mathrm{He},
\end{aligned} \tag{2.7}
$$

typically accompanied by neon-burning:

$$
\begin{aligned}
{}^{20}\mathrm{Ne} + {}^{4}\mathrm{He} &\rightarrow {}^{24}\mathrm{Mg} + \gamma \\
{}^{24}\mathrm{Mg} + {}^{4}\mathrm{He} &\rightarrow {}^{28}\mathrm{Si} + \gamma,
\end{aligned} \tag{2.8}
$$

and oxygen-burning:

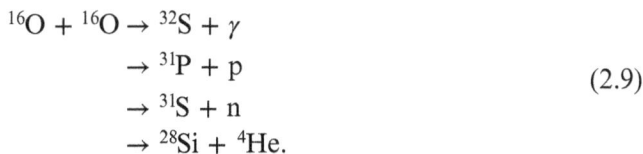

$$
\begin{aligned}
{}^{16}\mathrm{O} + {}^{16}\mathrm{O} &\rightarrow {}^{32}\mathrm{S} + \gamma \\
&\rightarrow {}^{31}\mathrm{P} + \mathrm{p} \\
&\rightarrow {}^{31}\mathrm{S} + \mathrm{n} \\
&\rightarrow {}^{28}\mathrm{Si} + {}^{4}\mathrm{He}.
\end{aligned} \tag{2.9}
$$

This pattern of competing reactions leading to different products is a common feature of nucleosynthesis involving relatively massive nuclei. The protons, neutrons and α particles released by such reactions rapidly undergo further processing. Because of its stability, the dominant product of this phase of stellar evolution is silicon.

The final phase in the life of a massive star leads to elements in the vicinity of the iron peak (figure 2.1). A logical route would be the direct association of two ${}^{28}\mathrm{Si}$ nuclei to form ${}^{56}\mathrm{Ni}$ (which then decays to ${}^{56}\mathrm{Fe}$); but in reality the situation is more complex. When temperatures above 2×10^{9} K are reached in the star's core, ambient thermal photons have sufficient energy to remove protons, neutrons and α particles from heavy nuclei, and these are rapidly captured by other nuclei to form a range of products. Destructive and constructive reactions operate in parallel, and the equilibrium abundance of any given element will depend on its binding energy. The binding energy per nucleon is greatest for elements in the region of Fe, and so their abundances build up in the core of the star.

The reactions producing elements up to Fe are predominantly exothermic: there is a net yield of energy because each successive compound nucleus is more tightly bound than its parent nuclei. However, the energy 'value' of each successive 'fuel' decreases with atomic mass. For example, the release of binding energy per nucleon for He burning to C is much less than for H burning to He: for this reason, stars spend most of their lives on the main sequence and successively less time in each subsequent cycle as they evolve through the giant or supergiant branches. After leaving the main sequence, a massive star will typically run through He, C, O and Si burning cycles in no more than a few million years. Because Fe has the greatest binding energy per nucleon, there are no exothermic reactions that can utilize it to form still heavier elements. When no further energy source is available, the core undergoes catastrophic collapse, resulting in a supernova explosion.

2.2.4 Element production in supernovae

Massive stars that end their lives as core-collapse supernovae are crucial to astrobiology, not only because they are a major source of several of the elements essential to life (figure 2.2) but also because they evolve quickly (<100 Ma from birth to death), and would therefore have been the first stars to produce and distribute *any* elements heavier than primordial H and He. After proceeding through each successive burning cycle described above, a star on the point of becoming a supernova is expected to have an onion-like structure, with an Fe-rich core surrounded by successive layers dominated by the ashes of each previous cycle; meanwhile, the outermost layers typically remain H-rich, displaying little evidence of the dramatic changes within. Core-collapse then generates a violent shock wave that propagates outward through this material, leading to further nuclear processing as well as driving dispersal of the products into the surrounding interstellar medium.

Far from peeling off in an orderly fashion, the layers surrounding the core are transformed into a soup of highly energized particles, in which many of the compound nuclei formed in earlier fusion cycles are split apart, generating copious supplies of α particles, protons and neutrons. Observations and models are in reasonable agreement on the abundances of the elements emerging from this melee. Again, destructive and constructive reactions operate in parallel, tending to favor the more stable products and thus accounting for both the odd–even pattern and the iron peak in the observed abundances (figure 2.1). One crucial area that remains somewhat uncertain is the extent to which C survives this turmoil, as it may readily be converted to O by capture of α particles (equation (2.6)) in the supernova, and the rate of this reaction is not precisely known. Material ejected from modern super-novae such as the well-studied SN 1987A is observed to be O-rich, and models for first-generation supernovae predict C:O mass abundance ratios substantially lower than the solar value of about 0.5 (Kobayashi *et al* 2006). This important point is discussed further in the next section. The availability of free neutrons also enables elements heavier than those at the iron peak to be formed: this occurs in a step-wise fashion, in which neutron capture (resulting in a unit increase in atomic mass) is followed by β-decay (emission of an electron, leading to a unit increase in atomic number). All of these products merge into the interstellar medium as the supernova remnant expands.

So far we have focused on stars that evolve independently of one another, i.e. single stars or those in wide binary or multiple systems. The evolution of stars in close, interacting systems is influenced by transfer of matter between members, a situation that can also induce a supernova explosion. Consider, for example, a close binary pair, neither of which has enough mass to become a core-collapse supernova. As they evolve, the more massive of the two becomes a white dwarf while the other is still a red giant. At this point, mass transfer from the red giant may cause the white dwarf to reach a critical mass and undergo thermal runaway. Such explosions are thought to make significant contributions to the production of several elements of intermediate mass (figure 2.2). Yet more dramatic possibilities include stellar mergers, such as the merging of two neutron stars, a potentially important source of heavy elements formed by neutron capture.

2.3 Distributing the products of nucleosynthesis

A simple schematic overview of galactic evolution is shown in figure 2.4. Stars are born when clumps of interstellar matter become dense enough to condense under their own gravity (see section 3.3). The first generations of stars would have been composed entirely of primordial elements (effectively just H and He). As stars evolve, they synthesize new elements by means of the processes described above, and return a portion of their mass to the interstellar medium: mass loss is effectively a feedback loop in this flow chart. Initially, mass loss takes the form of a stellar wind, i.e. a continuous, tenuous stream of ionized particles emanating from the stellar atmosphere. In general, this wind contains no new elements formed *within the star* as they remain locked deep inside (but see section 2.3.2 below for an important exception). Later, when the star reaches the final crisis point in its evolution, the core collapses and the outer layers are ejected. Elements falling into the collapsing cores are effectively lost from the cycle, but the ejected envelopes—a supernova remnant or a planetary nebula[5], depending on the mass of the star—carry products of nucleosynthesis into the interstellar medium, where they become available for inclusion in future generations of stars. This feedback loop is responsible for a

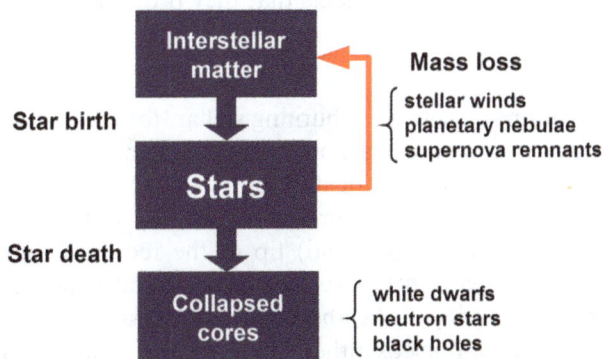

Figure 2.4. Schematic flow chart of galactic evolution. Interstellar matter gives birth to new stars. Stars synthesize new chemical elements, and return a portion of their mass to the interstellar medium, prior to ending their lives as collapsed stellar remnants.

---

[5] The term 'planetary nebula' is potentially misleading. In this case, it refers to the ejected outer layers of a star with insufficient mass to trigger a supernova explosion, so called because these objects sometimes resemble planets when viewed through small telescopes: the Ring Nebula (M57) is a well-known example. They should not be confused with protoplanetary disks, i.e. planet-forming disks around young stars.

gradual increase in *metallicity*—a term used by astronomers to represent the combined abundances of all elements heavier than H and He—from effectively zero to about 2% by mass over cosmic history to date.

2.3.1 Low and high mass stars: a cosmic case of 'The tortoise and the hare'

By cosmic standards, massive stars evolve so rapidly that their products begin to be recycled almost instantly. By the time our Galaxy had reached a mere 0.1% of its current age, supernovae were already detonating. Moreover, the manner of their evolution and explosive demise (section 2.2.4) ensures that their ejecta are enriched in heavy elements. Does it follow that slower-burning stars of lower mass are relatively minor contributors to the feedback process? This question is important because it has implications for the timescales over which life became possible, and the answer depends on a number of factors discussed in this section. A star 50% heavier than the Sun, for example, is predicted to live for 3–4 Ga, or about 25% of the age of the Galaxy, and the Sun itself for 10–11 Ga, or about 75%, so production and recycling of their products is significantly delayed in comparison to stars that end in core-collapse supernovae. An individual supernova event will typically eject a greater mass of material than a typical planetary nebula (1–5 $M_\odot$ compared with 0.1–1 $M_\odot$), but this is more than offset by the fact that low-mass stars are far more numerous. The number of stars born with a given mass (M) is found to be proportional to $M^{-2.35}$, an empirical result termed the *initial mass function*: to give examples, solar (1 $M_\odot$) stars outnumber 10 $M_\odot$ stars and 20 $M_\odot$ stars by factors of about 224:1 and 1140:1, respectively. So whereas massive stars are the sole contributors at early times, less massive stars are potentially major (and probably dominant) contributors over the longer term for the vital C, N, O elements that they produce.

2.3.2 Carbon stars

Of special interest amongst these slow-burning stellar 'tortoises' are those belonging to a class of red giants known as carbon stars. In general, O is found to be more abundant than C in the atmospheres of stars with well-studied spectra, typically by a factor of about two, as is the case for our Sun (figure 2.1). But in carbon stars, which lie near the low-temperature (right-hand) tip of the red giant branch in the HR diagram (figure 2.3), this pattern is reversed. Of course, it is no surprise that these stars would be manufacturing C by the triple-α process (section 2.2.3), but it is unusual for the products of nucleosynthesis to reach the surface of a star, where we are able to observe them. The most likely explanation arises from the fact that red grants contain deep *convective layers*.

Figure 2.5 compares models for the internal structure of the Sun and a red giant. In a main-sequence star like the Sun, the energy released by fusion in the core reaches the outer layers in two stages, initially by radiation (photons are continuously absorbed and re-emitted) and then by convection as hot gas is transported toward the surface. In the Sun, the convective layer is relatively shallow, about 15% of the radius of the star, but in an ageing red giant it may reach depths approaching 80%–90% of the star's radius. As energy production in the dying star begins to shut

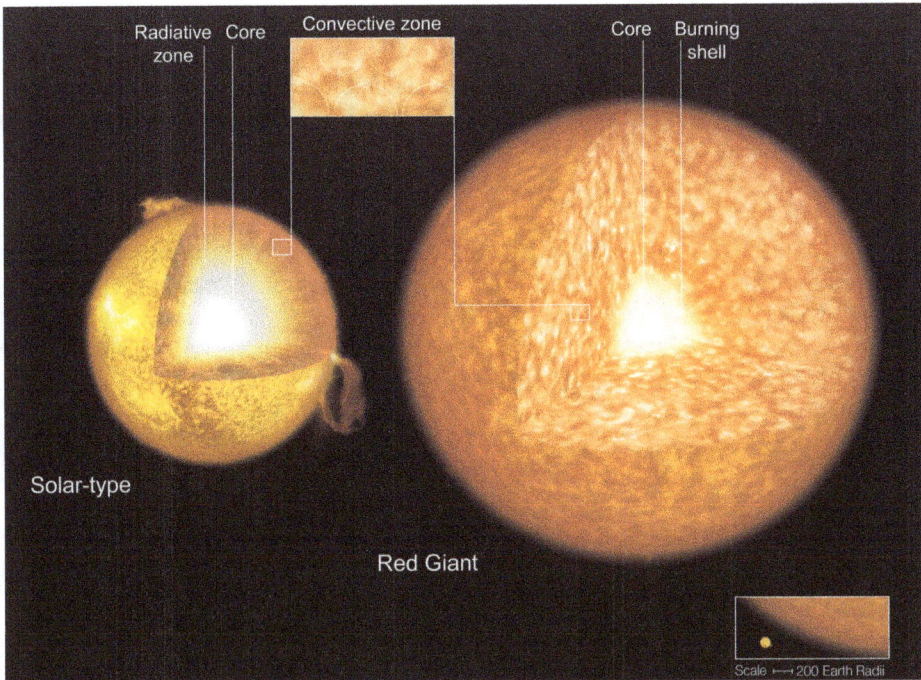

Figure 2.5. Comparison of models for the internal structure of the Sun and a red giant star, illustrating the major difference in the depths of the convective layers. The two images are not to scale (their relative size is shown by the inset at bottom right). Image credit: European Southern Observatory, annotated by the author.

down, it enters a phase of instability in which thermal pulsations may occur: loss of thermodynamic pressure causes the core to contract and heat up, which re-ignites the thermonuclear furnace, which causes expansion and cooling, which shuts down energy production, and the cycle repeats. During these pulsations, carbon reaches the bottom of the convective layer and is transported to the surface, a process referred to as 'dredge-up'. Nitrogen (produced by CNO-cycle H-burning) may also be dredged in this fashion.

Dredge-up causes the abundance of C in the star's atmosphere to rise steadily, and if it reaches a level exceeding that of O a dramatic change occurs in the composition of the stellar wind emanating from the star. As it expands and cools, physical conditions in the wind become suitable for dust particles to condense. In a wind with 'normal' abundances, i.e. with C/O < 1, the dust is composed mostly of silicates; but when C/O > 1 the excess C (remaining after CO molecules have formed) condenses to form carbonaceous macromolecules and dust, including polycyclic aromatic hydrocarbons and grains composed of amorphous (sooty) carbon, graphite and silicon carbide. This material becomes part of the star's outflow. Carbon stars subsequently generate C-rich planetary nebulae when their cores collapse to become white dwarfs, further contributing to the availability of C in the interstellar medium, and, ultimately, in the environments where future solar systems will be born.

Not all red giants evolve to become carbon stars. The probability appears to depend on initial metallicity—the extent to which the abundances of the heavier elements had already increased as a result of the feedback cycle (figure 2.4) prior to the birth of the star. Interestingly, carbon stars appear to be most common amongst stars of *lowest* initial metallicity (Mouhcine and Lancon 2003), as would have been the case at very early times in cosmic history.

In summary, these results suggest that a rich source of carbon feedback was instigated some 3 Ga after the birth of our Galaxy, when first-generation stars somewhat more massive than the Sun (1.5–3 $M_\odot$)—those massive enough to produce C but not to consume it—ended their lives as carbon stars. Whether this was a more important source of C than supernovae (section 2.2.4) remains an open question, but it seems likely to have been at least comparable.

2.3.3 Metallicity over time

The general increase of heavy-element abundances over time is confirmed by observations of metallic lines in the spectra of stars with reliable age estimates. Results are summarized in figure 2.6, which shows a plot of stellar age versus metallicity index[6] for various groups within our Galaxy. Those that have undergone significant dredge-up are excluded, so each star's spectrum provides a benchmark of the elemental abundances in the interstellar medium at the time that it formed. The sample includes stars from different locations as well as different age groups, and this also affects their metallicity (central regions of the Galaxy became enriched

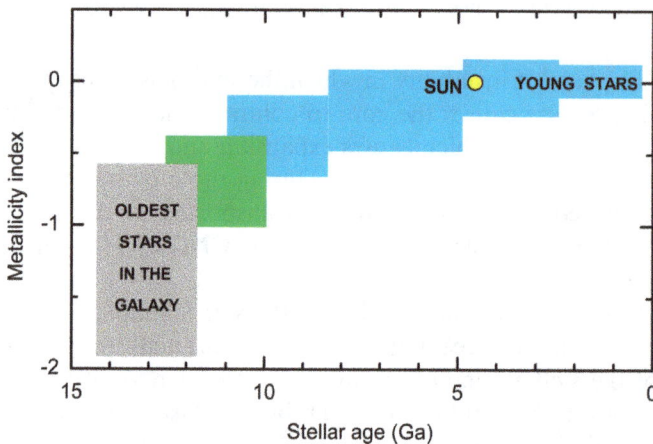

Figure 2.6. Plot of age versus metallicity index. The rectangles represent distributions for stars in different age groups and locations in our Galaxy: 'thin disk' (cyan), 'thick disk' (green) and 'halo' (grey); see Trimble (1997). The Sun is also shown for comparison (yellow circle). The trend from left to right illustrates the general increase in heavy-element abundances resulting from the cycle of star birth, nucleosynthesis and mass loss, over the lifetime of the Galaxy.

---

[6] The metallicity index is a measure of the Fe/H abundance ratio on a logarithmic scale, defined to be zero for solar abundances and positive or negative for stars more or less metal-rich than the Sun.

more rapidly than outer regions, and the halo contains older stars than the disk). Nevertheless, a general trend is evident in figure 2.6, indicating that the rate of enrichment was far more rapid when the Galaxy was young than it is today.

As noted at the outset of this chapter, the rise in heavy element abundances, as new generations of stars are born from the ashes of previous ones, is an essential step toward the formation of rocky planets and the origin of life. Observations confirm that the incidence of planets increases with stellar metallicity (see section 9.1.2): indeed, this is true of gas-giants as well as rocky planets, as expected if coagulation of solids in a protoplanetary disk is the starting point for all planet formation (section 4.2). Our Sun was born when the Galaxy was about 67% of its current age, by which time the average metallicity was not far short of the present level[7]. We return to this topic in chapter 9.

Questions and discussion topics

- Suggest any other elements (additional to those listed in table 2.1) that might have significant roles in terrestrial biology.
- Which are the *least* abundant of the elements most essential to life? Are there any in particular that might be limiting factors for life, in terms of availability in a typical planetary environment?
- Why do elements with even atomic numbers tend to be more abundant in the Universe compared with the odd-numbered ones?
- The Sun has been powered by H → He fusion throughout its lifetime to date, but when we analyze the Sun's atmosphere, we do not find unusually large amounts of He. Why not?
- The Sun will spend about 10 billion years fusing H into He, but only about 10 million years fusing He into heavier elements. What is the primary reason for this difference?
- What properties of a star determine whether it produces carbon dust or silicate dust during its red giant phase?
- This chapter is focused on single stars like the Sun, but it is estimated that at least 30% of all stars in our Galaxy are in binary systems. Consider what implications the presence of a close binary companion might have for (i) stellar evolution, and (ii) planetary habitability.
- Most cosmologists support the evolutionary (big bang) model for the origin of the Universe over the steady state model. What implications, if any, might there be for theories of the origin of life if the steady state model were proven to be correct?

---

[7] The Sun's metallicity appears to be fairly typical of its age group and location in the galactic disk. Suggestions that the Sun is anomalously metal-rich have appeared in the literature, but more recent results appear to rule this out (see Asplund *et al* 2009).

References and further reading

Asplund M, Grevesse N, Sauval A J and Scott P 2009 The chemical composition of the Sun *Ann. Rev. Astron. Astrophys.* **47** 481

Kobayashi C, Umeda H, Nomoto K, Tominaga N and Ohkubo T 2006 Galactic chemical evolution: Carbon through zinc *Astrophys. J.* **653** 1145

Marty B, Alexander C M O 'D and Raymond S N 2013 Primordial origins of Earth's carbon *Rev. Mineral. Geochem.* **75** 149

Mouhcine M and Lancon A 2003 Carbon star populations in systems with different metallicities: Statistics in local group galaxies *Mon. Not. R. Astron. Soc.* **338** 572

Trimble V 1997 Origin of the biologically important elements *Orig. Life Evol. Biosp.* **27** 3

IOP Concise Physics

Origins of Life
A cosmic perspective
Douglas Whittet

Chapter 3

Molecules in space: from interstellar clouds to protoplanetary disks

When our Milky Way Galaxy first formed some 14 Ga ago, it was entirely gaseous and composed of primordial hydrogen and helium. With the onset of star formation, the processes described in the previous chapter resulted in a gradual enrichment of the interstellar medium (ISM) with heavier elements. More than a merely passive medium that accumulates these elements until they are recycled into new stars, the ISM is now known to harbor regions conducive to a rich chemistry[1]. Production of the more abundant 'condensible' elements, C, O, Mg, Si and Fe, enabled the formation of solid particles (dust grains) in stellar ejecta, composed of metals, oxides, silicates, and solid carbon in its various allotropes. The presence of dust grains, in turn, facilitated chemical processes by providing a substrate for molecule formation. Thus, notwithstanding the disadvantages of low density and harsh physical conditions (extremes of temperature, irradiation, and shock waves), the ISM evolved to become a crucible for chemical evolution, some of the products of which can be found in the birth remnants of our own solar system. Significantly, the environments most conducive to chemical evolution, such as the dark cloud shown in figure 3.1, are those where new stars are born. The ISM is thus a vital agent in the cycle of galactic evolution that leads to planets and life. This chapter traces the path of interstellar chemistry and assesses its significance as a universal resource that delivers raw materials to potentially habitable planets in new solar systems.

---

[1] This realization was driven by major advances in astronomical spectroscopy at radio and infrared wavelengths during the second half of the 20th century, which enabled detection of a wide range of interstellar and circumstellar molecules, both in the gas and in or on solid particles, leading to the birth of the field of astrochemistry (see Herbst 2014 for a review). For a list of all detected molecules to date, see https://en.wikipedia.org/wiki/List_of_interstellar_and_circumstellar_molecules.

doi:10.1088/978-1-6817-4676-0ch3

Figure 3.1. A composite of visible and near-infrared images of Barnard 68, a representative dense molecular cloud that is expected to become a future site of star formation. The cloud is opaque to visible radiation because of absorption and scattering by dust: the distribution of background stars should be approximately uniform across the image if the dust were removed. Image credit: European Southern Observatory.

3.1 Chemistry in the interstellar medium

3.1.1 Interstellar environments

Much of interstellar space is by terrestrial standards a near-perfect vacuum, and thus seems an unpromising candidate for chemistry. To make the numbers more manageable, those who study it tend to express densities in terms of atoms per unit volume rather than mass per unit volume: the average density of interstellar gas in the disk of our Galaxy is about 1 atom cm^{-3}. Regions significantly denser than this meager average ($\gtrsim 10$ atoms cm^{-3}) are referred to generically as *clouds* (although they are not clouds in the terrestrial sense, being local accumulations rather than local condensations). The dominant constituent, gaseous hydrogen, can exist in three states: ionic, atomic or molecular. These states correspond to three different types of interstellar environment: intercloud (a tenuous plasma with densities typically well below the average, in which clouds are embedded), diffuse clouds (in which H is mostly atomic), and dense clouds (in which H is mostly molecular). In this context, 'dense' implies number densities $\gtrsim 100$ atoms (or molecules) cm^{-3}. The ISM approximates to an ideal gas in which pressure equilibrium is maintained between a cloud and the surrounding intercloud medium if the quantity nT (the product of number density and temperature) is constant: thus, the temperature of an interstellar environment varies inversely with its density. The tenuous intercloud medium is permeated by ionizing photons and extremely hot ($T \sim 5 \times 10^5$ K), whereas dense clouds (e.g. figure 3.1) are opaque to this radiation and extremely cold ($T \sim 15$ K). The intercloud medium accounts for most of the volume of the ISM, but the clouds account for most of its mass. That dense clouds are *molecular* clouds is understandable, as the rate of molecule formation increases with density, and the cold inner regions of the clouds are self-shielded from the harsh external radiation field that could otherwise destroy molecules as quickly as they are produced.

3.1.2 Surface chemistry

How do molecules form in these conditions? Even in relatively dense interstellar gas, where $n \sim 10^4$ cm^{-3}, converting atoms to molecules is not straightforward. When two atoms encounter each other, they must release binding energy in order to pair as a stable diatomic molecule. In a much denser gas such as a planetary atmosphere (which, in the case of the Earth at sea level, amounts to $n \sim 10^{19}$ cm^{-3}), this is facile because of the frequency of collisions—binding energy of the nascent molecule can be transferred almost instantaneously to kinetic energy of a third particle. But at interstellar densities, the probability of such a near-simultaneous 3-body encounter between gas atoms is vanishingly small. Instead, some molecules can lose binding energy by emitting a photon, but many of the observed interstellar molecules (including H_2, the most abundant of all) have no suitable transition, and so the molecule quickly dissociates:

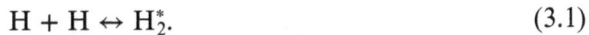

$$H + H \leftrightarrow H_2^*. \tag{3.1}$$

The asterisk denotes an unstable exited state, and the two-way arrow indicates that the reaction reverses. However, once the ISM had been sprinkled with stardust, as the result of stellar production and condensation of the relevant heavy elements, an effective new route to molecule formation became available:

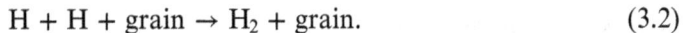

$$H + H + \text{grain} \rightarrow H_2 + \text{grain}. \tag{3.2}$$

This reaction does not require the simultaneous encounter of three particles, because an individual atom can become physically attached to the surface of the grain and remain there, if the temperature is low enough, until another atom is encountered (see figure 3.2 for a schematic overview of surface processes). The grain, which typically contains $\sim 10^9$ atoms, thus provides a meeting point as well as a heat sink that absorbs some or all of the binding energy, enabling the new molecule to reach a stable state. In the case of H_2, the molecule will generally desorb from the grain surface (some of its binding energy converting to translational energy); but molecules containing heavier elements are more likely to stick to the grain at ambient dense-cloud temperatures, resulting in the build-up of an icy surface coating. Given the ubiquity of H, and the status of O as the next most abundant element capable of forming chemical bonds, it is no surprise that these coatings

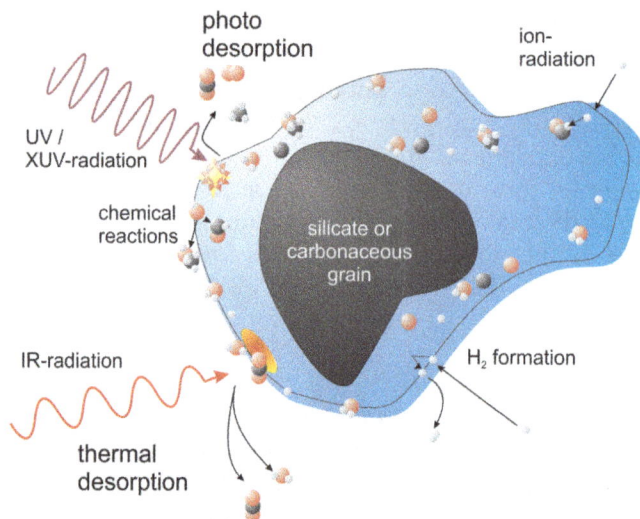

Figure 3.2. Schematic representation of processes occurring on the surface of an interstellar dust grain. A silicate or carbonaceous grain (shaded dark gray) typically forms in a stellar outflow. Subsequently, inside a molecular cloud, it acquires an icy mantle (shaded blue) as O and other species attaching to the surface undergo hydrogenation. Surface recombination of H is the primary source of H_2 in the ISM. Subsequent star formation drives photochemical reactions, and thermal and photo-desorption of the ices. Image courtesy of Helmut Zacharias, from Siemer *et al* (2014), reproduced with permission of The Royal Society of Chemistry.

(mantles) are typically observed to contain as much as 70% H_2O, which forms easily by a simple two-step process:

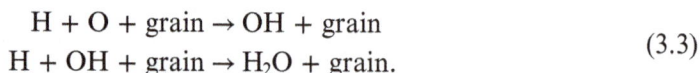

$$H + O + grain \rightarrow OH + grain$$
$$H + OH + grain \rightarrow H_2O + grain. \tag{3.3}$$

Again, the grain provides the substrate and heat sink, absorbing binding energy released by each step.

3.1.3 Gas-phase chemistry

Once H_2 is available in the interstellar gas, new routes to molecule formation are enabled that do not depend on the presence of a substrate. Exothermic *chemical exchange reactions*, in which two reactants produce two (or more) products, are energetically favored because the excess is transferred to kinetic energy of the products. As an example, consider the reaction:

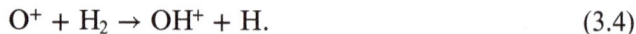

$$O^+ + H_2 \rightarrow OH^+ + H. \tag{3.4}$$

An H atom has been exchanged between the reactants, and, in contrast to reaction (3.1) above, this one is balanced by the formation of two products. Another point of note is the fact that one of the reactants is ionized, bearing a positive charge owing to prior loss of an electron. This is an important factor in astrochemistry because the electric field of the ion polarizes the charge of the neutral reactant, leading to a net Coulomb attraction. As a result, such ion-molecule reactions are more efficient than would be the case for equivalent neutral reactions. Even within dark clouds such as Barnard 68 (figure 3.1), where ionizing photons from the external radiation field do not penetrate, a degree of ionization is maintained by energy from cosmic rays[2].

Reaction sequences based on this principle can explain many of the observed interstellar molecules, including some relevant to astrobiology. For example, reaction (3.4) initiates a route to water via the sequence

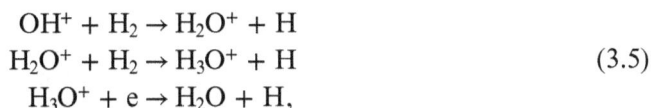

$$OH^+ + H_2 \rightarrow H_2O^+ + H$$
$$H_2O^+ + H_2 \rightarrow H_3O^+ + H \quad\quad (3.5)$$
$$H_3O^+ + e \rightarrow H_2O + H,$$

and an exactly analogous sequence starting with N^+ and involving one additional step produces NH_3. Carbon hydrogenation is initiated in a different manner because the equivalent first step $(C^+ + H_2 \rightarrow CH^+ + H)$ is endothermic and therefore inhibited in interstellar clouds. A more likely route requires prior formation of the highly reactive H_3^+ ion:

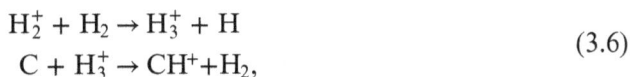

$$H_2^+ + H_2 \rightarrow H_3^+ + H$$
$$C + H_3^+ \rightarrow CH^+ + H_2, \quad\quad (3.6)$$

with further H-additions leading to CH_4. Carbon also undergoes oxidation to produce CO (the most abundant interstellar molecule after H_2), which can form in a number of ways, including by the ion-molecule reaction

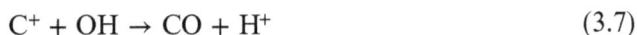

$$C^+ + OH \rightarrow CO + H^+ \quad\quad (3.7)$$

and the neutral reaction

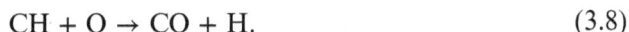

$$CH + O \rightarrow CO + H. \quad\quad (3.8)$$

These and a host of other reactions have been used in chemical models that predict abundances in accord with the observations for the majority of known species.

3.1.4 Deuterium fractionation as a diagnostic of interstellar chemistry

Approximately one in every 50 thousand hydrogen atoms in the interstellar gas is deuterated, i.e. its nucleus contains a neutron as well as a proton, this level of abundance being determined primarily by element production in the early Universe (section 2.2.2). The factor 2 mass difference between a normal H atom and deuterium (D) can lead to very significant increases in *molecular* D/H ratios compared with this general average for atomic gas. In principle, any H-bearing molecule can be deuterated by inclusion of D in place of H. Both surface chemistry and gas-phase chemistry are affected. Atoms that become physically adsorbed onto a surface are restrained by weak electrostatic forces, and are subject to escape by sublimation at a rate that increases rapidly with temperature. The probability that

---

[2] Cosmic rays are highly energetic particles (protons and other atomic nuclei) that travel through space at relativistic speeds, thought to originate primarily in supernova explosions (section 2.2.4).

any given atom will pair with another to form a molecule (section 3.1.2) depends on its residence time on the grain. At a given temperature T, the kinetic energy of the particle is

$$\tfrac{1}{2}\,mv^2 = \tfrac{3}{2}\,kT = \text{constant}, \tag{3.9}$$

so if mass m is doubled (by substituting D for H), the mean speed v is reduced by a factor $\sqrt{2}$. This reduction in mean atomic speed increases the mean residence time for D relative to H, and hence its probability of molecule formation. Because of this, the abundance of HD relative to H_2 formed on dust grains is substantially higher than the average for atomic interstellar gas, and this bias is carried forward into other hydrogenated molecules that form in the gas as well.

Gas-phase chemistry (section 3.1.3) tends to amplify the fractionation. In particular, the important H_3^+ ion (equation (3.6)) is readily deuterated by the reaction

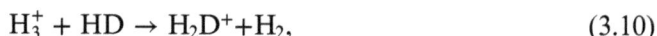

$$H_3^+ + HD \rightarrow H_2D^+ + H_2, \tag{3.10}$$

which proceeds efficiently in the forward direction (and not the reverse) at molecular-cloud temperatures. Consequently, molecules that form from it also bear this isotopic signature. Some species become enhanced by factors ~ 100 or more over the average for atomic gas. Hence, the D/H ratio is a helpful diagnostic of interstellar chemistry, providing, for example, evidence for the presence of presolar interstellar matter in carbonaceous meteorites (section 4.3.2), remnants from the birth of our own solar system.

3.2 The rise of molecular complexity

The processes described in the previous section generate a wide variety of interstellar molecules. In the gas phase, these range from simple diatomic and triatomic species such as H_2, CO, CH, OH, CN, HCN and H_2O to carbon-chain molecules containing as many as 13 atoms. Also present are polycyclic aromatic hydrocarbons (PAHs), a class of large ($\gtrsim$ 30-atom) carbonaceous molecules composed of interconnected benzine rings, with various shapes and sizes, probably originating for the most part in the outflows of carbon stars (section 2.3.2). The solid phase consists of icy mantles frozen onto silicate or carbon dust grains (figure 3.2), composed of H_2O with an admixture of other species such as CO, CO_2, NH_3, CH_3OH and CH_4, some formed *in situ* on the surface, some captured from the gas. In general, 'small' molecules (i.e. those containing fewer atoms) are far more abundant than large ones, reflecting the nature of the formation process in which molecules are built step-wise by addition of atoms: the larger the molecule the greater the number of steps, each with an associated probability. There are limits to the degree of chemical complexity that can be reached by such processes in the cold inner regions of a dense (prestellar) cloud, where only exothermic reactions are generally possible. New routes toward greater complexity become available once stars begin to form, creating internal sources of energy within the cloud. In this section, we consider steps toward complexity in the cold environments of prestellar clouds; these determine the initial conditions for chemistry driven by the onset of star formation, discussed in the next section.

3.2.1 Organic interstellar molecules and the search for glycine

The majority of known interstellar molecules are organic in nature[3]. The preponderance of organic molecules is especially great for the larger ones: indeed, all of those detected in the gas phase to date containing seven or more atoms are organic, a clear demonstration of the ubiquity and versatility of carbon chemistry. Some of the detected species are exotic by terrestrial standards, such as various radicals and ions that would quickly react to form other species in a planetary environment. Others are familiar and directly implicated in terrestrial biology. Examples of the latter include formaldehyde (H_2CO, a precursor to sugar), formic acid (HCOOH), acetic acid (CH_3COOH), methylamine (NH_2CH_3), and of course ethyl alcohol (CH_3CH_2OH). For the most part, the network of chemical reactions that produces these species is reasonably well-understood: see Herbst and van Dishoeck (2009) for a review.

It is notable that a chemical network capable of producing molecules such as methylamine and formic acid has reached a level of complexity where it is just a step away from glycine (NH_2CH_2COOH), the simplest amino acid. It is thus reasonable to suppose that glycine is present at some (low) level of abundance in the ISM, and concerted efforts have been made to detect it, a quest sometimes regarded as the holy grail of astrochemistry. Detections have been claimed, but so far none have held up to scrutiny. The problem is a practical one, arising from the multitude of vibrational and rotational energy levels that occur in a molecule as complex as glycine. Because of this, the predicted radio spectrum is extremely complicated and it is easy to obtain an accidental coincidence with an unidentified spectral line, leading to a false-positive result. Only by detecting many predicted lines over a wide range of frequencies and with the expected relative intensities can a detection in the gas phase be confirmed. The solid-state spectrum predicted for glycine embedded in ice on dust is far simpler (because rotational transitions are suppressed in solids), but in this case detection is hindered by ambiguity: the vibrational spectra of different organic molecules are often very similar. In summary, the observations do not exclude the possibility that glycine is produced in molecular clouds, but these problems have so far prevented confirmation. Spectroscopic searches for other molecules of comparable or greater complexity face the same problem.

3.2.2 Hydrogenation versus oxidation: a vital branching point for surface chemistry

Laboratory experiments on realistic analogs of interstellar ice mixtures show that complex organic molecules, including amino acids, may be produced, even at low temperature, when the ices are exposed to energetic photon or particle irradiation.

---

[3] There is no precise, universally-agreed definition of the term 'organic'. In general, an organic molecule contains carbon and at least one other element from the set H, N, O, but with certain exceptions such as the oxides CO and CO_2, and mineral carbonates.

However, the degree of molecular complexity depends rather critically on the assumed initial composition, in particular the abundance of hydrogenated forms of carbon relative to fully oxidized carbon. Ice mixtures in which C resides primarily in CO_2 give much lower yields than those rich in hydrogenated species such as CH_4, H_2CO and CH_3OH. The composition of the ices in a dense cloud just prior to the onset of star formation is determined by an important branching point in grain-surface chemistry with respect to carbon monoxide (CO), by far the most abundant C-bearing species produced in the gas phase (section 3.1.3). When CO attaches to a grain, it may subsequently undergo hydrogenation:

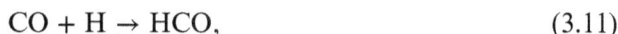

$$CO + H \rightarrow HCO, \qquad (3.11)$$

or oxidation:

$$CO + OH \rightarrow CO_2 + H. \qquad (3.12)$$

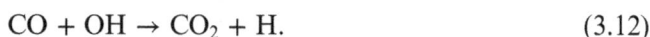

The HCO radical produced by reaction (3.11) may readily undergo further H-addition surface reactions to produce formaldehyde (H_2CO) and methanol (CH_3OH), two molecules that may in turn be processed to more complex species when an energy source becomes available. In contrast, CO_2 produced by reaction (3.12) presents an obstacle to further processing because of its inherent stability. The situation is analogous to that discussed in section 1.4.1 for the terrestrial atmosphere.

Disappointingly from an astrobiological perspective, available data suggest that CO_2 is typically more abundant than hydrogenated forms of carbon in interstellar ices, indicating that the oxidation route is generally favored. But there are important exceptions: certain regions have been identified in which CH_3OH, in particular, is a factor of 3–4 times more abundant than expected relative to this general trend. The cause of these apparent 'sweetspots' for chemical evolution is not fully understood, but a key factor may be the availability of atomic O. In the early phases of a cloud's evolution, O is readily available, both H_2O and CO_2 form efficiently by reactions (3.3) and (3.12), and any HCO that forms by reaction (3.11) may also then be oxidized to $CO_2 + H$ rather than hydrogenated to H_2CO. At a later time in the evolution of the cloud, most of the O is already tied up in molecules such as H_2O, CO, CO_2 and O_2, and the hydrogenation route is then favored. This may coincide with the formation of dense prestellar cores inside the cloud, a time of rapid freeze-out of CO onto grains. See Whittet et al (2011) for detailed discussion.

3.3 Protostars and the chemical heritage of protoplanetary disks

3.3.1 A brief primer on star formation

A star begins to form when a prestellar core within a molecular cloud reaches a critical mass and density, such that gravitational forces overcome thermal gas pressure, at which point it begins to undergo gravitational collapse. This condition may arise randomly, or it may be triggered by a specific event such as a collision between clouds or a shock wave from a nearby supernova. Stars do not generally form in isolation. A molecular cloud will usually undergo fragmentation as it collapses to produce a family of stellar siblings that become members of a star cluster. The fragmentation process spawns both individual stars and binary or multiple systems. Depending on the initial mass of the cloud, the cluster may contain anything from ~10 to tens of thousands of members. As discussed later, the number of stars in the group has a profound effect on the environment experienced by each individual member during the early phases of its evolution, when planetary systems are being forged. An example of a populous region of star formation is shown in figure 3.3.

As an individual protostar collapses, the release of gravitational energy causes its central regions to heat up, and at this point it becomes a copious emitter of infrared radiation. Initially, gravitational heating is moderated by this radiative loss, until the protostar becomes sufficiently dense that infrared photons are mostly absorbed. Ultimately, as collapse continues, the internal temperature rises until it is sufficient for nuclear reactions to be ignited in the core of the new star, and it then becomes

Figure 3.3. Hubble Space Telescope image of the star-formation region NGC 602 in the Small Magellanic Cloud, a satellite Galaxy of the Milky Way. Luminous stars in the central cluster are gradually dissipating the natal molecular cloud, while others are still forming within the cone-shaped structures above, below and to the right of center. Image credit: HST/NASA/ESA.

luminous at visible wavelengths. Further gravitational collapse is prevented by thermodynamic pressure, and the star evolves into the main sequence (section 2.2.3).

Each prestellar cloud fragment that gives birth to a star will have some net rotation. The effect of rotation will not be apparent initially, as gas motions at low densities are governed primarily by turbulence. But as the cloud collapses, conservation of angular momentum results in flatting of the envelope into a rotating protoplanetary disk centered on the protostar, as shown schematically in figure 3.4. Initially, the disk is dusty and opaque, containing ~0.1 μm-sized particles that absorb and scatter starlight efficiently. In the polar regions above and below the plane of the disk, however, the star's radiation and stellar wind emerge unobstructed, generating a bipolar outflow. The disk itself gradually becomes transparent

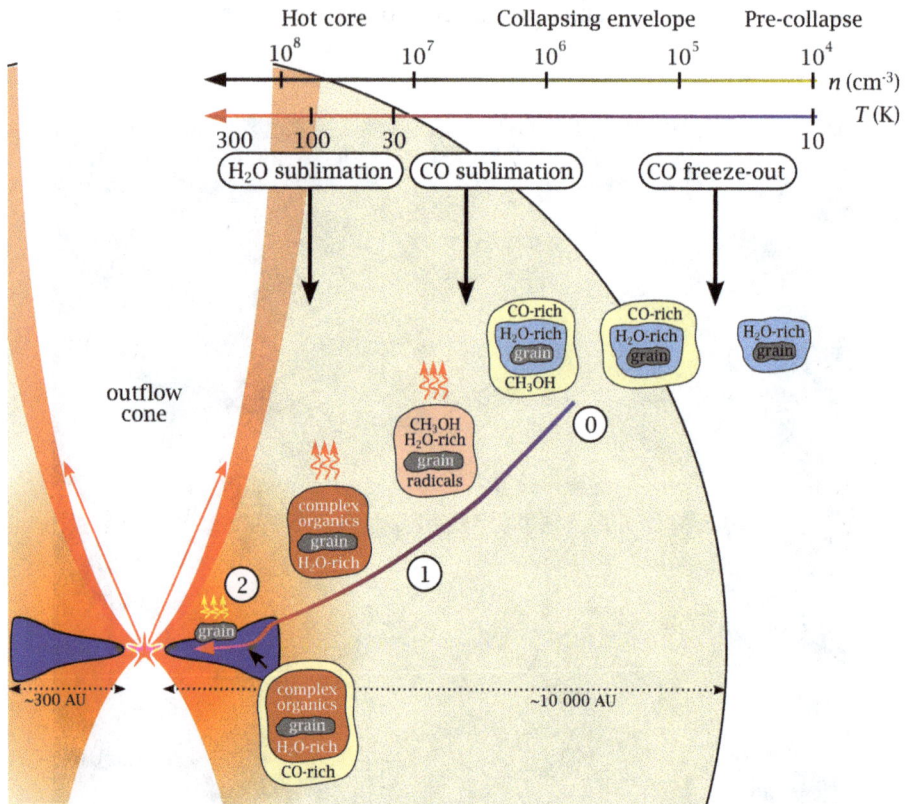

Figure 3.4. Schematic representation of a protostellar envelope, illustrating the range of environments from pre-collapse core to collapsing envelope (beige), hot core (red) and protoplanetary disk (blue). Temperature and density scales refer to the envelope, not to the disk. Dimensions are given in astronomical units (the mean Sun–Earth distance; 1 AU = 1.496 × 10^{11} km). The initial composition (labeled 0) represents material inherited from the parent cloud; primary and secondary processing occurs in the collapsing envelope (1) and hot core (2). The status of solid grains and their mantles is illustrated at various points (not to scale!). Grain dimensions are typically ~0.1 μm. Image credit: E van Dishoeck and R Visser, adapted from Herbst and van Dishoeck (2009).

during the planet-building process, as many small particles are aggregated into fewer large ones (section 4.2).

The basic process described here is thought to be qualitatively similar for single stars over a range of masses, albeit on differing timescales. Evolution from initial collapse to the main sequence may take some 10–20 Ma for the Sun, compared with only 1–2 Ma for a 10 $M_\odot$ star. The origin of our Sun and solar system is discussed further in chapter 4.

3.3.2 Chemical evolution in a protostellar envelope

Chemical evolution in a region of active star formation is driven by changes in physical environment imposed by the star-formation process. These changes affect both matter falling into a collapsing protostar and ambient interstellar material exposed to the radiation and winds from luminous stars that have already formed in the region (figure 3.3). The first of these situations is discussed here, the other in the final section.

Figure 3.4 charts the evolution of material during the formation of an individual protostar. The composition of the material accreting from the parent cloud (stage 0 in figure 3.4) is determined by processes outlined in the previous sections. Initially at molecular-cloud temperatures ($\lesssim 15$ K), this material is gradually warmed and ices begin to sublimate (stage 1). CO, the most volatile of the common ices and the last to freeze out during the prestellar phase, is the first to sublimate as temperatures exceed ~17 K. However, H_2O, the most abundant of all the ices, is retained until temperatures exceed ~100 K. Other species, such as NH_3 and CH_3OH, that formed simultaneously with (and became embedded in) the H_2O–ice may be retained to temperatures well above their nominal sublimation points. As the ices warm up, the individual molecules become more mobile. Ices formed in the parent cloud are initially amorphous, i.e. the molecules are randomly ordered in the solid, but warming may lead to crystallization as they move into more favorable energy states. Mobility also enhances the opportunities of chemical reactions, including some with significant activation energies that were not possible in the parent cloud. Models suggest that much of the chemistry leading to the synthesis of complex molecules occurs between radicals on the surfaces of these warming particles in the collapsing envelope.

In regions closer to the protostar, termed *hot cores* (shaded red in figure 3.4), temperatures reach levels sufficient to sublimate H_2O[4]. Other species that were previously trapped in the ice are also released at this time: gaseous molecules such as NH_3 and CH_3OH are often detected in hot cores in concentrations much higher than can be accounted for by gas-phase chemistry but are readily explained by surface chemistry followed by sublimation. Temperatures in hot cores range from 100 to 300 K, and densities are typically three or four orders of magnitude higher than in the parent cloud—conditions highly conducive to further chemical evolution (stage 2 in figure 3.4). Products include a number of species of potential astrobiological

---

[4] Hot cores around solar-mass stars are sometimes termed *hot corinos*.

significance detected *only* in hot cores (and not in cold clouds), such as formamide (NH_2CHO), aminoacetonitrile (NH_2CH_2CN), methyl mercaptan (CH_3SH) and acetone (CH_3COCH_3).

Another likely product of this phase is a class of carbonaceous material termed organic refractory matter (also known as tholin or kerogen). As the name implies, this material remains stable at higher temperatures than most organic molecules. It is broadly similar in composition to insoluble organic matter found in terrestrial sedimentary rocks, characterized by high molecular weight ($\gtrsim 1000$) and a non-specific composition, typically including many randomly-linked aromatic and aliphatic hydrocarbons and other subunits. Laboratory experiments show that organic refractory matter is produced when interstellar ice analogs are exposed to irradiation and heating. This class of material is found to be a common constituent of carbonaceous meteorites, indicating that it was formed during the early evolution of our solar system.

Hot cores are exposed to radiation from the protostar, but material close to the disk, which accumulates in the equatorial plane of the system (shaded blue in figure 3.4), are progressively shielded from this radiation by the opacity of the disk itself, resulting in a temperature gradient. In contrast to the warm inner regions, temperatures toward the outer region edge of the disk may revert to values as low as those in the parent cloud (but now at much higher density)—cold enough not only to retain any ices that escaped sublimation but also to refreeze volatiles such as CO from the gas. These disks become the feeding grounds for planetesimals, and ultimately, planets (section 4.2), as the ice-coated dust grains collide, coagulate and grow.

Once a star has ignited its nuclear fuels and become self-luminous, the surrounding envelope is gradually dispersed. This process begins toward the polar regions of the system, where the stellar wind emerges unhindered by the disk (figure 3.4). Dispersal of the disk is much more gradual: in warm regions close to the star, solids are limited to refractory materials such as metallic iron and silicates, but beyond a certain radius termed the *frost line* (also known as the snow line), ices and other volatiles, including many of the organic molecules formed by processes described in this chapter, are available for inclusion into planetary bodies. The frost line for H_2O in the protoplanetary disk of our solar system was located at a radial distance of about 2.7 AU from the Sun, between the present-day orbits of Mars and Jupiter.

3.3.3 The significance of the birth environment

The energy output of a new star has the potential to affect not only its own envelope but also those of its siblings and any remaining fragments of the birth cloud that have yet to be converted into stars. Whether the effect is significant depends on the stellar population density (how closely they are spaced) and, particularly, on the mass range of the stars involved. High-mass stars are by far the most significant in this context because of their luminosity, their strong stellar winds, and the speed at which they evolve. Stars of mass 10 $M_\odot$ or more may reach the main sequence while their lower-mass siblings are still forming: examples can be seen in figure 3.3. Indeed, it is possible for the most massive stars ($\gtrsim 25$ M$_\odot$) to evolve through their entire life-cycles and become supernovae before star formation in other regions of the same birth cloud is complete.

Some potential consequences of this interaction have implications for life, and they may be either positive or negative. Compression of ambient interstellar material caused by the impact of an energetic stellar wind or supernova shockwave may trigger further episodes of star formation. Being much hotter than solar-mass stars (figure 2.3), massive stars radiate intensely at shorter wavelengths, predominantly in the ultraviolet. Both the radiation they emit and the stream of energetic particles in their winds may erode the envelopes of sibling stars, potentially limiting the sizes of their protoplanetary disks. However, ultraviolet radiation can also promote photo-chemistry, yielding additional opportunities for organic synthesis.

An interesting hypothesis that has been discussed in this context concerns a possible trigger for homochirality—specifically the preference, in the case of terrestrial biology, for left-handed forms of the chiral amino acids (see section 1.1). The radiation emanating from young stars is often polarized, as a result of the interaction of photons with dust grains in the presence of magnetic fields. In addition to linear polarization, the radiation may include a circular component, in which the electric vector of the electromagnetic wave rotates in a circle normal to the direction of propagation. Laboratory studies show that when chiral molecules are exposed to circularly polarized light, the degree of absorption depends on the handedness, i.e. the direction in which the electric vector rotates. If an assemblage of chiral molecules that is initially racemic (i.e. containing equal numbers of L and D forms) is exposed to circularly polarized *ultraviolet* radiation that is potentially energetic enough to dissociated the molecules, the balance of L and D is altered because one form is dissociated faster than the other; which out of L and D survives best depends on the handedness of the radiation. The radiation emerging in a given direction from the dusty envelope of a massive star has an equal probability of being left-handed or right-handed, but it is possible for a neighboring protostar to be exposed preferentially to just one, depending on its location. This hypothesis has been suggested as a possible cause of the excess of L-amino acids detected in carbonaceous meteorites, discussed in the next chapter.

Whether a specific star-formation region is subject to such developments depends critically on its mass. To illustrate this, consider two contrasting birth environments: (i) a quiescent, low-mass cloud (the Taurus region of our Galaxy contains a well-studied example), and (ii) a giant molecular cloud (the closest one to our Sun lies in the Orion 'sword' region). Quiescent clouds are often filamentary rather than centrally condensed, and give birth to at most a few hundred stars. Giant molecular clouds are more compact and spawn stars in much greater numbers, typically several thousand or more (e.g. figure 3.3). But in either case, the statistics governing the relative numbers of stars that form *in a given mass range* is the same (see section 2.3.1). Because of this, the probability that a quiescent cloud will produce even one star of mass $\gtrsim 10$ M$_\odot$ is very low, whereas a giant molecular cloud is likely to produce several. So in summary, stars forming in a quiescent cloud will evolve in comparative isolation, with little or no interaction between the siblings; but those forming in a giant molecular cloud will experience much stronger ultraviolet radiation fields and winds, and perhaps even a local supernova event.

Which of these environments gave birth to our own Sun and solar system? This question is addressed in the next chapter.

Questions and discussion topics

- In what fundamental ways do chemical processes in an interstellar cloud differ from chemical processes in a planetary atmosphere?
- Explain the importance of ion–molecule reactions in interstellar chemistry.
- Why is deuterium enrichment a signature of interstellar chemistry?
- Why is even the simplest amino acid, glycine, very difficult to detect in the interstellar medium?
- If glycine is confirmed as an interstellar molecule by future research, do you expect that this discovery will be regarded as a major advance in our quest to understand the origins of life?

References and further reading

Boogert A C A, Gerakines P A and Whittet D C B 2015 Observations of the icy Universe *Ann. Rev. Astron. Astrophys.* **53** 541

Fukue T *et al* 2010 Extended high circular polarization in the Orion massive star forming region: Implications for the origin of homochirality in the Solar System *Orig. Life Evol. Biosph.* **40** 335

Herbst E 2014 Three milieux for interstellar chemistry: Gas, dust, and ice *Phys. Chem. Chem. Phys.* **16** 3344

Herbst E and van Dishoeck E F 2009 Complex organic interstellar molecules *Ann. Rev. Astron. Astrophys.* **47** 427

Öberg K I *et al* 2011 The Spitzer ice legacy: Ice evolution from cores to protostars *Astrophys. J.* **740** 109

Pontoppidan K *et al* 2014 Volatiles in protoplanetary disks *Protostars and Planets VI* (University of Arizona Press) p 363 http://dx.doi.org/10.2458/azu_uapress_9780816531240-ch016

Siemer B *et al* 2014 Free-electron laser induced processes in thin molecular ice *Faraday Discuss.* **168** 553

Whittet D C B *et al* 2011 Observational constraints on methanol production in interstellar and preplanetary ices *Astrophys. J.* **742** 28

Origins of Life
A cosmic perspective
Douglas Whittet

Chapter 4

The origin and evolution of our solar system

Previous chapters reviewed the origin and distribution of the chemical elements, and traced the evolution of biologically relevant compounds—water, organic molecules, and planet-building dust and ices—from interstellar clouds into protostellar envelopes and protoplanetary disks. We next focus on our own solar system, the only planetary system so far known to host life. The ultimate goal is to attempt to determine what characteristics led to this outcome and whether they are universal: is our solar system special, or is it merely a typical example of a Galaxy-wide phenomenon? Progress toward answering this question requires, in the first instance, an objective assessment of the nature, origin and evolution of the solar system itself, which is the principal goal of this chapter. The next step is to make detailed comparisons between our own and other planetary systems, a topic addressed in chapter 9.

4.1 The Sun's birth environment

Hypotheses that attempt to explain the origin of our Sun and planetary system have been developed over many years: the nebular hypothesis proposed by Immanuel Kant in the 18th century is a direct precedent of modern theories that account for the origins of both our own and other systems in terms of a rotating disk centered on a protostar that condensed from an interstellar cloud. Until relatively recently, however, it has been customary to assume that most stars form in isolation, given that their separations in later life are typically vast compared with their individual sizes (close binaries being obvious exceptions). We now know that this is not necessarily a sound assumption, even for single stars.

The significance of the birth environment is summarized in the previous chapter (section 3.3.3). The most important statistic that determines the outcome in a specific case is the population census of the group: this, in combination with the initial mass function (section 2.3.1), determines the probability that the group contains high-mass stars. The transition between a quiescent region (lacking high-mass stars) and a more populous region (containing at least a few) occurs at about 100–200 members. In the former case, the assumption that members evolve in isolation is reasonable; in the latter it is not. Consequences for a star born in a heavily populated region include the possibility of close encounters with other, more massive, members, leading to gravitational disruption of the protoplanetary disk, exposure to harsh radiation fields, energetic winds and shock waves, and infusion with material ejected by supernovae.

What of our Sun? As a mature, isolated star, it is now evidently far removed from its birth environment. Statistics on current star formation in the Galaxy do not help to resolve the question, as they suggest that the *a priori* probability of a solar-mass star forming in a group of more than or less than the critical size is about the same. Nevertheless, clues remain that enable us to make an informed assessment, summarized here.

- *The structure and size of the disk:* The orderly nature of the planets' present-day orbits around the Sun (concentric, almost circular, almost coplanar, and with a preferred direction of motion) argues against major disruption of the protoplanetary disk resulting from a close encounter with a massive neighbor. However, some minor disruption is consistent with the eccentric orbits of smaller, more distant objects such as Sedna. The distribution of mass declines rapidly at solar distances beyond the current orbit of Neptune: estimates of the total for dwarf planets and asteroids in the outer solar system amount to no more than a few percent of the mass of the Earth. This suggests that the protoplanetary disk may have been limited to a radius of about 40 AU, perhaps by photo-evaporation or erosion.

- *Short-lived radioactive isotopes:* The presence of radioactive isotopes in the protoplanetary disk that have since decayed may be inferred by detection of their daughter products. Studies of meteorites have identified several examples, such as ^{26}Al and ^{60}Fe, which decay to ^{26}Mg and ^{60}Ni with half-lives of 0.6 and 1.5 Ma, respectively—timescales shorter than the protostellar phase of a solar-mass star. Interpretation of this finding is still somewhat controversial; but by general consensus, the most likely explanation is that these isotopes were injected by one or more contemporary supernova explosions. This is a reasonably probable outcome if the Sun formed in a large group, an extremely unlikely coincidence if it did not.

On balance, these arguments support the conclusion that the Sun formed in a large group: detailed statistical analyses suggest one with a membership of a few thousand. An even larger group could have resulted in catastrophic disruption of the protoplanetary disk, a smaller one would have reduced the probability of supernovae to explain the isotopic evidence. See Adams (2010) for a detailed review.

Large groups of newly-formed stars often evolve to become gravitationally-bound star clusters, after the remaining placental material from the birth cloud has been dispersed. Yet the Sun is clearly not now a member of a cluster—if it were located, say, in the Pleiades, we should expect to see dozens of stars brighter than Sirius in the nighttime sky! Current isolation by no means contradicts the evidence for a populous birth environment, however, as all but the densest star clusters disperse on timescales shorter than the age of the Sun. The random motions of individual members occasionally result in speeds sufficient for escape, and as the number of members dwindles the net gravitational field of the cluster declines, loosening its grip on those that remain.

4.2 The solar nebula and the origin of the planets

4.2.1 Overview

In broad terms, the physics of star formation is reasonably well understood (section 3.3.1). The hypothesis of a rotating protoplanetary disk (the solar nebula) centered on the proto-Sun provides a robust model of the origin of our planetary system, consistent with most of its present-day geometric and kinematic properties. These include its flatness (the inclinations of the orbits of the major planets differ by no more than a few degrees) and its preferred direction of orbital motion and spin: the planets orbit the Sun in the counterclockwise (prograde) direction as viewed from north, which is also the direction of the Sun's rotation, characteristics assumed to be inherited from the original disk. One caveat to the model is a prediction that the Sun should carry most of the angular momentum of the system, inconsistent with its leisurely current rate of rotation (the period is about 25 days). In the past, this problem led investigators to consider alternative theories for the origin of the solar system, such as a tidal interaction with a passing star, but these face far greater problems. It is now widely accepted that the solar nebula hypothesis is broadly correct and that the Sun's implied rapid initial rate of rotation has since slowed, most likely as the result of magnetic and/or frictional interactions with the disk and envelope during the bipolar outflow phase of its early evolution (section 3.3.1). This conclusion is greatly strengthened by the fact that we can observe protoplanetary disks orbiting young solar-mass stars in current regions of star formation elsewhere in the Galaxy: these are, in effect, analogs of the early solar system.

4.2.2 Frost and soot lines

A striking characteristic of our planetary system that must be taken into account is the clear division between the inner region, occupied by relatively small, rocky planets, and the outer region, occupied by giant planets and their icy moons (see figures 4.1 and 4.2). It is probably no coincidence that the frost line for H_2O (section 3.3.2 and figure 4.3) lies in the annulus separating these two zones. Inside the frost line, solids available to coalesce into planets were limited by ambient temperatures to the more refractory materials, such as metals and rock-forming minerals, leading naturally to planets with bulk compositions similar to that of the Earth, with its iron-rich core and silicate-rich mantle and crust. Beyond the frost line, H_2O–ice was added to the mix, along with (at somewhat greater distances) solids containing the more volatile ice-forming molecules such as CO_2, CH_4, N_2, and NH_3. This 'dirty snowball' mix of dust and ices is the likely starting point for the assembly of icy planetesimals that we now recognize as comets. Some of these grew into substantial

Figure 4.1. Portrait gallery of our planetary system. The eight major planets are displayed in order of increasing distance from the Sun from left to right; relative sizes are to scale but distances are not. Important characteristics of the system that distinguish it from many exoplanetary systems include (i) spatial segregation of the rocky planets (Mercury, Venus, Earth, Mars) from the gas/ice giants (Jupiter, Saturn, Uranus, Neptune), confined to the inner and outer regions, respectively (see figure 4.2), and (ii) the absence of any planets intermediate in size between these two groups (Neptune, the smallest of the giants, is about four times larger than the Earth). Image credit: NASA/JPL.

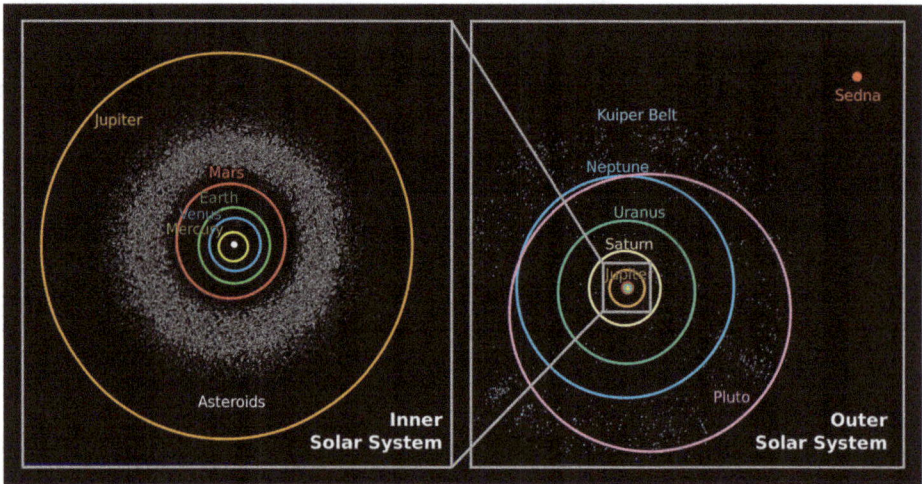

Figure 4.2. Distribution of planetary orbits in the inner and outer regions of the solar system. The loci of the main asteroid belt, the Kuiper belt, the dwarf planet Pluto and the distant minor planet Sedna are also shown (the latter at its approximate perihelion distance). Image credit: NASA/JPL, with minor edits by the author.

icy worlds, such as the giant-planet moons and the dwarf planet Pluto. The largest may have become the cores of giant planets (section 4.2.4).

The distribution of carbon compounds in this preplanetary mix is harder to determine because C can form such a wide variety of solids. The possibilities range from refractory crystals (diamond, graphite) that could survive in the inner regions (but were probably not abundant), to volatiles such as CO, CO_2, CH_4 and other organic species that condense in the outer disk. Intermediate between these extremes

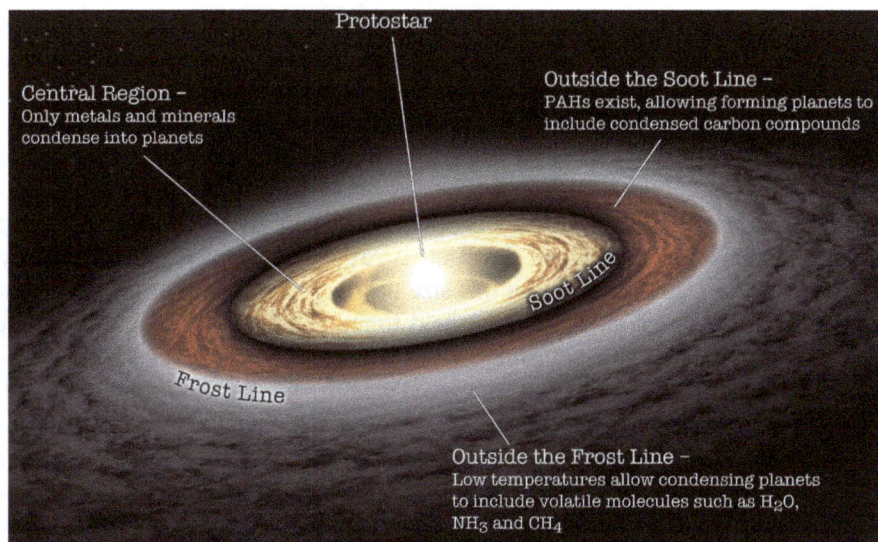

Figure 4.3. Schematic representation of the solar nebula, illustrating the loci of annular zones inside which temperatures are too high for solid H_2O and polycyclic aromatic hydrocarbons to be available (the frost line and soot line, located at approximate radii of 2.7 AU and 2.0 AU, respectively). Image credit: Invader Xan (supernovacondensate.net) and NASA/JPL.

are polycyclic aromatic hydrocarbons (PAHs; section 3.2) and organic refractory matter (section 3.3.2). In particular, the ubiquity of PAHs in the interstellar medium, where they account for some 10%–20% of the available C, implies that they should also be common in protoplanetary disks. Calculations suggest that PAHs should remain stable at solar distances $\gtrsim 2$ AU, corresponding to the 'soot line' in figure 4.3. This is corroborated by observations of C-rich asteroids in the main belt, and by laboratory work showing that the meteorites thought to originate from them contain both PAHs and organic refractory matter.

4.2.3 Accretion: from dust to planets

The planet-building process in protoplanetary disks is an active area of research. Some of the details are not yet fully worked out, but there is general consensus on the physical mechanisms and sequence of events that led to the growth of planets in the solar nebula. Initially, solids in the disk are mostly dust grains with sizes $\lesssim 1$ μm. The general rotation of the disk is approximately Keplerian, i.e. the constituent particles should ideally orbit the Sun according to Kepler's 3rd law of planetary motion[1]. Departures from perfect Keplerian rotation arise as gas and dust particles interact with each other. The gas is partially supported by its own pressure and therefore rotates at a slightly slower rate, leading to size-dependent frictional drag on the dust. Collisions between the dust particles themselves are frequent, and if they are gentle enough they may be constructive, resulting in growth by accretion. Aggregate

---

[1] Kepler's 3rd law states that $P^2 \propto a^3$, where P is the orbital period and a is the semi-major axis.

particles are initially held together by weak electrostatic forces, and thus tend to be fragile. However, the most stable aggregates tend to grow at a rate proportional to their cross-sectional area: the larger they become, the more particles they sweep up. This process builds aggregates ranging up to about a meter in size.

The growth of meter-sized aggregates into kilometer-sized planetesimals is the least well understood phase of the process. As they grow, the aggregates acquire increasingly large speeds relative to other particles in their vicinity, raising the probability that collisions will be destructive rather than constructive. However, once a few aggregates break through this size barrier, rapid growth can be resumed. Planetesimals are massive enough to be held together by self-gravity and therefore less vulnerable to destructive collisions. Their gravitational fields are also sufficient to influence the motions of neighboring particles in the disk, leading to a rate of growth that depends on mass; so again, the largest grow the most rapidly. This phase continues until the disk is cleared out. Over timescales of a few million years, the overall transition is from vast numbers of small particles to a small number of large ones, each with bulk composition reflecting the radial zone in the disk where it formed. The overall scale of the process is illustrated by a simple calculation: the number of 1 μm-sized dust grains required to build a planet the size of the Earth is $\sim 10^{38}$.

Evidence of past accretion can be seen on different size scales in the present-day solar system. Firstly, carbonaceous chondrites, the most pristine remnants of the process that we can study in the laboratory, have a highly granular structure. The grains are predominantly silicates, with sizes typically in the range 0.1 μm to 1 mm, infused with organic material—as expected if they accreted in what is now the asteroid belt (see section 4.3.2). Secondly, when we observe geologically unmodified surfaces throughout the solar system, from Mercury to the Earth's Moon, from Mars to the icy moons of the giant planets, we see that they are pockmarked by impact craters—scars of the late stages of accretion.

4.2.4 Origins of the giant planets

The accretion process described above suggests a route to bodies with solid surfaces —rocky planets in the inner region, icy worlds in the outer region—but what of the gas giants? Although not themselves considered likely hosts for life, understanding their origins is nevertheless important in the astrobiological context because they may be highly influential in determining the habitability of a planetary system.

One possibility is that they simply represent a more extreme outcome of the same accretion process. The range of solids available for accretion is greatest beyond the frost line, but this is moderated by a systematic decline in density with radial distance toward the outer edges of the disk. So the zone of the disk lying beyond— but not far beyond—the frost line seems likely to be a sweetspot for growing large planetesimals, and it is suggestive that the most massive planet, Jupiter, now occupies this zone. A key difference, comparing giant planets with smaller ones, is that giants capture (and retain) the abundant light elements H and He. This is shown in figure 4.4, which plots escape velocity versus temperature, illustrating the ability

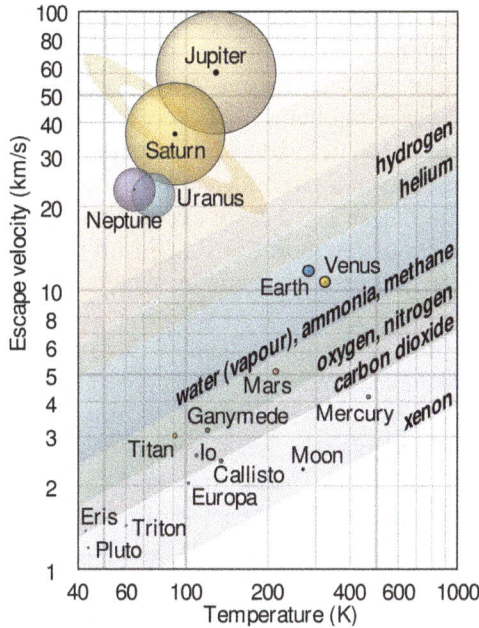

Figure 4.4. Plot of escape velocity versus temperature for planets and large moons in our solar system, with their sizes shown to scale. The diagonal shading indicates retention limits for various atmospheric gases with respect to leakage over cosmic timescales caused by kinetic motion. A given gas is retained in the atmosphere of a body with an escape velocity greater than the limiting velocity for the gas at the appropriate temperature. Image credit: Cmglee/svg (Wikimedia Commons).

of planetary bodies in the present-day solar system to retain various atmospheric gases. The escape velocity of a body increases with mass[2], and if it is massive enough to *retain* a particular gas at a particular temperature it is also able to *accumulate* it from the surrounding medium. Hence, a protoplanet that grows beyond a certain critical mass may acquire an atmosphere rich in H and He. The four giants are the only planets in our solar system that reached this critical mass.

A point in favor of the accretion model is that it explains the observed correlation in extrasolar planetary systems between the occurrence of giant planets and the metallicity of the host star (sections 2.3.3 and 9.1.2). At first sight, this correlation might seem counter-intuitive, given that planets such as Jupiter and Saturn are composed primarily of H and He rather than heavier elements; but in the accretion theory, they are 'seeded' by large planetary embryos composed of rock and ice that subsequently accumulated deep atmospheres, and most of the mass of these embryos is in elements such as C, N, O, Mg, Si and Fe that would be scarce in a low-metallicity system. Nevertheless, some giant planets do exist in low-metallicity systems, and these are hard to explain by accretion. Another potential objection to the accretion model is that the estimated time to build a planet as large as Jupiter could be 10 Ma or more, comparable with the timescale for dissipation of the disk.

---

[2] For a body of mass M and radius R, the escape velocity is $v_{esc} = \sqrt{2GM/R}$.

An alternative hypothesis proposes that giant planets are effectively failed stars rather than overgrown planets: in this view, they originate as instabilities in the protoplanetary disk, in a manner analogous to mechanisms proposed for the origin of binary star systems. The combination of forces arising from gas pressure and gravity in the rotating disk leads to local instabilities that cause clumps to form, and if they are massive enough the clumps condense to become giant planets. If this is the case, they might form quite quickly, before accretion has had time to build large planetesimals. Another important point is that instabilities can arise in a purely gaseous disk, thus providing a possible route to planet formation in low-metallicity systems.

It seems reasonable to suppose that either of these mechanisms might operate according to local circumstances. Which of them gives the more realistic account of the birth of giant planets in our solar system is not yet known. The powerful gravitational fields of these planets have the potential to impart catastrophic disruption on their siblings, but our local ones now occupy stable, nearly-circular orbits that confine them to a safe distance from the terrestrial planet zone—a situation not replicated in many exoplanetary systems (see section 9.1 for further discussion). This circumstance was almost certainly crucial to the emergence of life on Earth, and we need a better general understanding giant-planet formation in order to assess how it arose and whether it is typical.

4.3 Time capsules from the early solar system

4.3.1 Comets and asteroids

The Sun's radiation and solar wind eventually cleared the disk of any remaining gas and dust that was not consumed by the planet-building process, leaving a system composed of bodies that range in size from the familiar major planets and their moons to dwarf planets, comets, asteroids and smaller debris. Traditionally, comets and asteroids have been considered as distinct groups, but both are in effect leftover planetesimals, composed mostly of dust, rock and ice in proportions that reflect their locus of formation: the comets predominantly icy, the (main-belt) asteroids predominantly rocky. As such, they provide unique samples of the birth aggregate.

There are three main reservoirs of these bodies in the present-day solar system: the main asteroid belt (between Mars and Jupiter), the Kuiper asteroid belt (beyond Neptune), and the Oort cloud. Whereas the two belts are toroidal and approximately coplanar with the planetary orbits, the Oort cloud is a vast spheroidal halo of comets that extends far beyond the confines of the original protoplanetary disk. If neighboring stars have such halos it is possible that they overlap and intermingle. The Oort cloud is thought to be populated, at least in part, by icy planetesimals that formed in the giant-planet zone and were subsequently scattered by near-miss encounters with those planets. Long-period comets such as Hale–Bopp, which passed through the inner solar system in 1997, are thought to be visitors from the Oort cloud.

Comets and asteroids are studied by both remote observation and laboratory analyses of physical samples. When a comet approaches the Sun the ice begins to sublimate, producing the familiar coma and tail, and releasing dust grains into interplanetary space. Cometary dust regularly enters the Earth's atmosphere, where samples may be collected by aircraft, but the value of these random samples is limited by the fact that the comet from which they came cannot be identified. However, Stardust, a targeted space mission, collected samples of dust from a specific comet, Wild 2, returning them to Earth in 2006. Wild 2 is thought to have originated in the outer disk; it was deflected in more recent times into an orbit with a perihelion close to that of Mars, where the rendezvous occurred. Stardust captured particles released from the comet by solar heating (it was not designed to collect icy material from the solid surface, a far more hazardous task). The mineral dust was found to consist of high-temperature condensates from the inner regions of the disk (rather than pre-existing interstellar grains), indicating that radial mixing of these particles into the outer disk must have occurred prior to aggregation of the comet. Traces of organic material were also found, and a notable result from the mission was the first detection of a cometary amino acid (glycine). See Elsila *et al* (2009) and Brownlee (2014) for detailed discussions of the results from Stardust.

4.3.2 Meteorites

Most meteorites are thought to be fragments of main-belt asteroids, resulting from collisions that deflected them into Earth-crossing orbits. On entering the Earth's atmosphere, typically at relative speeds of 10–20 km s^{-1}, they undergo intense frictional heating, accompanied by deceleration due to air braking. Typically, the smallest are completely ablated, but larger fragments (pebble-size and up) may reach the surface intact, their outer layers fused by melting, their interiors relatively unaltered. Meteorites may be grouped by composition into three broad categories (with considerable overlap)—metallic, stony, and carbonaceous—and these categories can be matched spectroscopically to observed classes of asteroid. Some stony meteorites resemble igneous rocks. Igneous and metallic meteorites appear to be fragments of planetesimals that grew large enough to undergo substantial internal heating. In contrast, carbonaceous chondrites (figure 4.5) are granular in structure and far richer in volatiles, containing both organic compounds and H_2O or OH groups in hydrated minerals. They appear to be fragments of relatively unaltered asteroids that formed mostly in the zone between the soot and frost lines in the solar nebula (figure 4.3).

The significance of carbonaceous chondrites as time capsules that enhance our understanding of the birth and early evolution of our planetary system is a recurrent theme in this book. Examples include:

- Definitive constraints on the age of the solar system (section 1.3).
- Data on the relative abundances of chemical elements in the solar nebula (section 2.1).
- Evidence for short-lived radioactive isotopes in the birth aggregate that indicate probable infusion of supernova ejecta (section 4.1).
- Insight into the nature of the accretion process in the formation zone (section 4.2.3).

Figure 4.5. The Allende meteorite, a carbonaceous chondrite that fell to Earth in 1969, delivering samples of material that formed in the solar nebula 4.56 Ga ago. The meteorite broke up in the atmosphere prior to impact: a fragment 1.4 cm across is shown (left). Also shown (right) is a cross-section through another fragment, revealing characteristic grainy internal structure. It includes ~1 mm-sized chondrules (glassy spherical grains) and lighter, irregularly-shaped Ca–Al-rich particles ranging up to ~1 cm in size, all embedded in a fine-grained (~1 μm) matrix of silicates mixed with carbon and organic matter. The carbonaceous material gives the meteorite its characteristic dark color. Image credits: James St John (Wikimedia Commons).

- A demonstration of principle for delivery of prebiotic molecules and water to Earth from space (sections 1.4.1 and 5.3.2), and a possible trigger for homochirality (section 3.3.3).

The remainder of this section focuses on the carbonaceous content of these meteorites, which typically contributes ~2%–3% of their weight, with a view to assessing its origins and potential significance as a resource for life.

Carbonaceous chondrites contain some individual grains shown, on the basis of isotopic analysis, to be survivors from the presolar birth cloud. Composed of graphite, diamond and silicon carbide, they appear to have originated in C-rich stellar outflows (section 2.3.2) or supernovae that predated the birth of the Sun. However, these presolar grains account for only a tiny fraction of the meteoritic carbon, most of which consists of organic molecules that were formed or reprocessed locally (in the solar nebula or on the asteroidal parent body). About two-thirds of this material is organic refractory matter, most likely the result of radiative processing of presolar ices (section 3.3.2), the rest is a complex mixture of soluble organics, including amino acids, alcohols, nucleic acid bases and sugars. Both the refractory and non-refractory organics show evidence of deuterium fractionation (section 3.1.4), at a somewhat lower level than is typical of interstellar molecules, suggesting that they contain a blend of presolar material and molecules formed at higher temperatures in the solar nebula.

The degree of heating experienced by an asteroid is an important factor that influences the extent to which the materials it inherited from the solar nebula underwent further *in situ* processing. There is naturally a gradient in solar heating across the asteroid belt, from the inner region populated mainly by igneous bodies to the central and outer regions where the parent bodies of the carbonaceous chondrites are found. Mineralogical studies show that liquid water must have been present at one time on these parent bodies, enabling chemical reactions in aqueous solution: thus, they were not merely accumulations of protoplanetary material but sites of chemical activity, modifying and extending the inventory of organic molecules that might subsequently be delivered to the Earth. Calculations show that solar radiation alone was insufficient to create aqueous conditions in the outer asteroid belt. Other energy sources might have included decay of radioactive isotopes, especially ^{26}Al, and induction heating caused by the passage of the asteroid through the ambient magnetic field. If decay of ^{26}Al was the dominant source, it is interesting to note that it might not have been available but for the intervention of a nearby supernova (section 4.1).

The amino acids extracted from carbonaceous chondrites exhibit a bias in favor of the left-handed forms of these chiral molecules, consistent with the chiral preference of terrestrial life: see Pizzarello (2016) for a review. The degree of bias varies from meteorite to meteorite and can be as high as 60%. This finding naturally raises the question of whether terrestrial contamination might be affecting the results, but this possibility is ruled out for two primary reasons: (i) the meteorites contain a wide variety of amino acids, with no particular preference for the 20 used by terrestrial biology; and (ii) the abundances of certain isotopes, particularly deuterium, are inconsistent with a terrestrial origin.

To date, this is the only known case of chiral asymmetry in organic molecules formed outside of the Earth's biosphere. The cause is unknown. A correlation has been found between the magnitude of the asymmetry and the degree of aqueous processing of the parent body, and it is known from experimental work that polymerization reactions on certain crystalline mineral surfaces can be chirally selective (see section 6.1.2). Nevertheless, it is difficult to formulate a satisfactory model based on processes that are entirely internal to the solar system because there is no obvious way to break symmetry on a large enough scale: for example, chemical reactions on one surface might favor L over D, but those on a neighboring surface are just as likely to favor D over L. The problem may be alleviated by invoking an external factor, such as exposure of the solar birth cloud to circularly polarized radiation of a given handedness (section 3.3.3). Any small asymmetry introduced in this way might then have been amplified by autocatalytic reactions on the asteroid and, indeed, upon subsequent delivery to Earth.

4.4 The evolution of habitability

The three fundamental ingredients needed for a world to create and sustain life as we know it are liquid water, an energy source, and an appropriate mix of prebiotic compounds (section 1.4.1). In this section, an adequate supply of prebiotic compounds from exogenous delivery and/or endogenous production is assumed; instead we focus primarily on the requirement for liquid water, as it relates to the energy sources available in the solar system. Of these, solar radiation is generally dominant, but the internal heat of a body generated by factors such as tidal friction and radioactivity also come into play.

4.4.1 The circumstellar habitable zone

The circumstellar habitable zone, often referred to as the Goldilocks zone (section 1.5) is defined as the range of orbital radii around a star within which the ambient temperatures on a planetary surface can support liquid water, given sufficient atmospheric pressure. A schematic illustration is shown in figure 4.6, comparing the Sun with stars of different luminosity. The discussion in this section is focused on the Sun but equally applicable to other stars, considered in chapter 9.

The mean equilibrium temperature T_e for a spherical planet of radius R in a circular orbit of radius r around a star of luminosity L can be estimated from the balance of radiant energy absorbed and emitted:

$$(1 - A)(\pi R^2)(L/4\pi r^2) = 4(\pi R^2)\varepsilon\sigma T_e^4, \tag{4.1}$$

where A is the planet's albedo (the fraction of incident energy that is reflected back into space), ε is its infrared emissivity (the efficiency with which heat energy is re-radiated), and σ is the Stefan–Boltzmann constant. As incoming and outgoing radiation both vary with the cross-sectional area of the planet (πR^2), T_e is independent of its radius:

$$T_e = [(1 - A)L/(16\pi\varepsilon\sigma r^2)]^{1/4}. \tag{4.2}$$

Using values of $A \approx 0.3$ and $\varepsilon \approx 0.6$ for the present-day Earth, together with the mean solar distance ($r = 1$ AU $= 1.496 \times 10^{11}$ m) and current solar luminosity ($L_\odot = 3.826 \times 10^{26}$ W), a temperature of about 289 K is predicted, reasonably consistent with current mean global temperatures.

Retaining the above values for A, ε and L while allowing r to vary in equation (4.2) provides an estimate of the range of orbital radii over which we might expect

Figure 4.6. Schematic comparison of the circumstellar habitable zones around three main-sequence stars of different luminosity: the Sun (center) and stars more and less luminous than the Sun (top and bottom). Habitable zones are shaded green; regions too hot and too cold to support liquid water on an earthlike planet are shaded red and blue, respectively. Image credit: NASA/JPL.

water to remain in liquid form on a hypothetical earthlike planet at different solar distances: T_e in the range 373–273 K corresponds to 0.6–1.1 AU. However, this simple calculation takes no account of changes in the planet's atmosphere that would occur under such circumstances, especially the release of greenhouse gases driven by increased evaporation of the oceans and destruction of biomass with rising temperatures, or changes in albedo resulting from variations in average cloud cover and the size of the polar ice caps. Sophisticated climate models have been developed that take these and other factors into account. There is a considerable spread in results, but reasonable estimates place the Goldilocks zone of the present-day solar system in the range 0.9–1.3 AU. The Earth is (for the time being) safely inside. Venus ($r \approx 0.7$ AU), having suffered a runaway greenhouse effect, is far too hot; Mars ($r \approx 1.5$ AU) is too cold.

These calculations assume the current solar luminosity, but this is not constant. Main-sequence stars such as the Sun are in a relatively stable state, producing sufficient energy by H → He fusion to maintain the balance between thermodynamic pressure and gravity (section 2.2.3); but significant changes occur, nevertheless, over cosmic timescales. As the abundance of He in the Sun's core gradually builds up, the density increases, leading to greater pressure; this causes an increase in energy production and, hence, luminosity. Since it became a stable main-sequence star some 4.56 Ga ago, the solar luminosity has increased by approximately 30%. It will continue to increase in the future, at a gradually accelerating rate, until the Sun evolves to become a red giant some 5 Ga from now (figure 4.7).

The steady increase in solar luminosity results in a corresponding outward migration of the habitable zone. If the other parameters in equation (4.2) remain the same, r increases in proportion with $\sqrt{L}$. This allows a prediction that at early times, soon after the birth of the solar system, the inner and outer radii of the habitable zone were smaller than their current values by a factor of about 0.84, placing the Earth near its outer edge. This is an important point that must be taken into account when assessing conditions leading to life on the early Earth, discussed in the next chapter. Extrapolating into the future, it is estimated that the Earth will remain in the habitable zone for perhaps another ~2 Ga, a much shorter time than the remaining main-sequence tenure of the Sun (Rushby *et al* 2013).

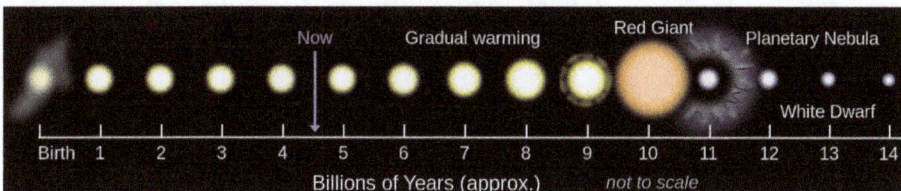

Figure 4.7. Schematic illustration of the Sun's life cycle, from birth and main-sequence evolution (yellow) to the red giant and white dwarf phases. The disk sizes are not to scale. The Sun is currently about halfway thought the main-sequence phase, during which it undergoes a slow but steady increase in luminosity. Image credit: Tablizer (Wikimedia Commons).

4.4.2 Energy from within

Bodies such as Venus that lie inside the inner edge of the Goldilocks zone, resulting in a runaway greenhouse effect, can probably be dismissed as potential hosts for life. But those that lie on the cold side, beyond its outer edge, seem far more promising. They do not necessarily lack liquid water because another source of heat may be available.

Rocky bodies accreting inside the frost line of the solar nebula have internal heat from the formation process, arising not only from the initial temperature of the material but also from kinetic energy released by collisions and gravitation energy released as denser material sinks toward their cores. This is augmented by heat generated by the decay of radioactive elements within, each according to its inherent timescale. The heat capacity of a planet is proportional to its volume, i.e. to R^3, whereas, over time, a planet loses internal heat at a rate proportional to its surface area, i.e. to R^2. It follows that a large planet retains internal heat (and thereby sustains volcanic activity) over far longer timescales than a small one, as illustrated in figure 4.8 for bodies in the inner solar system. To the best of our knowledge, the Earth is the only member of this group that remains volcanically active 4.5 Ga after birth.

A volcanically active planet is far better equipped to sustain aqueous environments in a locus where solar heating alone is insufficient. Local melting occurs when sources of geothermal heat come into direct contact with surface ice, allowing the possibility to generate hot springs or hydrothermal systems. More generally, volcanic emissions supply greenhouse gases to the atmosphere, enabling a planet to retain the energy it receives from the Sun more efficiently. If a planet has tectonic

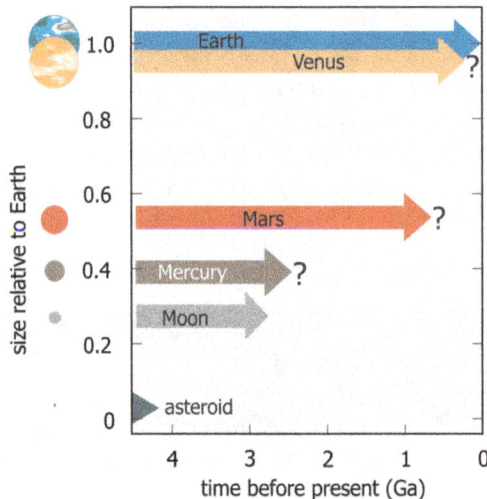

Figure 4.8. Chronology of volcanism related to size for the Moon and planets in the inner solar system. A representative igneous asteroid is also included. The length of the horizontal arrow for each body indicates the timescale over which it has maintained volcanic activity. Image credit: Bruce Watson, adapted from McSween (1994).

as well as volcanic activity, as in the case of the Earth, this allows greenhouse gases that might otherwise be trapped in the crust to be recycled to the atmosphere. Finally, the gravitational and magnetic fields of a planet govern the longevity of the atmosphere against leakage into space. The comparison between Earth and Mars is particularly interesting in this context. Substantially larger than Mars, the Earth has remained geologically active and held down a substantial atmosphere, thus maintaining aqueous surface environments over most of its history. Indeed, this size factor is probably just as vital to the habitability of the Earth as its locus in the Goldilocks zone. In contrast, Mars evidently supported surface water in the past but now appears dry, a situation arising from the combination of atmospheric leakage and decline in geologic activity, both resulting from its smaller size. See chapter 7 for a more detailed assessment of Mars as a potential host for life.

Turning to the outer solar system, the focus is on the icy moons of the giant planets. Some of these are of planetary size (the largest, Ganymede, Callisto and Titan, are intermediate in size between Mercury and Mars); but only Titan has a substantial atmosphere and none have surface conditions capable of sustaining liquid water. Nevertheless, several of these moons display evidence of surface activity, most notably the active 'ice volcanoes' observed on Saturn's moon Enceladus. It is probable that Enceladus and other icy worlds have subsurface oceans, maintained by heat from tidal friction and flexing imparted by gravitational interactions with the planet and other moons in the system. Jupiter's moon Europa is considered a prime candidate, with an ocean potentially much more voluminous than the Earth's. We return to this topic in chapter 8.

Questions and discussion topics

- Why would a nearby supernova explosion be less likely to have occurred during the very early history of the solar system if it was born in a small stellar group rather than in a large one?
- Explain why conservation of angular momentum leads us to expect that the Sun should carry most of the angular momentum of the solar system.
- Is it surprising that some meteorites are composed of igneous rocks?
- How can we be sure that amino acids detected in meteorites are not terrestrial contaminants?
- Why does the size of a rocky planet affect its level of geologic activity?
- Consider how the presence or absence of a magnetic field might affect the habitability of a planet.
- What arguments might be put forward to support or refute a hypothesis that early Venus was hospitable to life?

References and further reading

Adams F C 2010 The birth environment of the Solar System *Ann. Rev. Astron. Astrophys.* **48** 47

Brownlee D 2014 The Stardust Mission: Analyzing samples from the edge of the Solar System *Ann. Rev. Earth Planet. Sci.* **42** 179

Chiang E and Youdin A N 2010 Forming planetesimals in solar and extrasolar nebulae *Ann. Rev. Earth Planet. Sci.* **38** 493

Elsila J E, Glavin D P and Dworkin J P 2009 Cometary glycine detected in samples returned by Stardust *Meteoritics Planetary Sci.* **44** 1323

Güdel M 2007 The Sun in time: Activity and environment *Living Rev. Solar Phys.* **4** 3

Kress M E, Tielens A and Frenklach M 2010 The 'soot line': Destruction of presolar polycyclic aromatic hydrocarbons in the terrestrial planet-forming region of disks *Adv. Space Res.* **46** 44

Lecar M, Podolak M, Sasselov D and Chiang E 2006 On the location of the snow line in a protoplanetary disk *Astrophys. J.* **640** 1115

McSween H Y 1994 What we have learned about Mars from SNC meteorites *Meteoritics Planetary Sci.* **29** 757

Pizzarello S 2016 Molecular asymmetry in prebiotic chemistry: An account from meteorites *Life* **6** 18

Rushby A J, Claire M W, Osborn H and Watson A J 2013 Habitable zone lifetimes of exoplanets around main sequence stars *Astrobiology* **13** 833

van Dishoeck E F, Bergin E A, Lis D C and Lunine J I 2014 Water: From clouds to planets *Protostars and Planets VI* (University of Arizona Press) p 835

Origins of Life
A cosmic perspective
Douglas Whittet

Chapter 5

The early Earth: forging an environment for life

The Earth was born some 4.56 billion years ago, the largest of four planetary siblings that accumulated inside the frost line of the Sun's protoplanetary disk, as discussed in the previous chapter. Initial conditions set by the formation process—its location in the Goldilocks zone, its acquisition of enough mass to retain a substantial atmosphere—were crucial in determining the Earth's potential to host life. How was this potential realized? Was it happenstance or the inevitable outcome of favorable initial conditions that led to the emergence of life within the first billion years of the Earth's history? We do not yet have definitive answers to these questions, but we may begin to address them by reviewing what is known of evolving conditions on the early Earth. External influences must be assessed, including the solar luminosity, the threat of planet-sterilizing impacts, and contributions to the organic/volatile inventories of the planet from infalling material. Advances in our understanding of conditions during this early epoch are made as investigators develop new and sophisticated techniques to extract all possible information from the oldest available samples of the Earth's crust.

5.1 The Earth–Moon system

The Earth is unusual in having a relatively large moon: the only case amongst the major planets in our solar system of a moon greater than 5% of the planetary size, and the only large moon to accompany any planet inside the frost line. Indeed, one might with some justification consider the Earth–Moon system as a binary planet. Intermediate in size between Mercury and Pluto, the Moon would perhaps be a borderline case for 'major planet' status if it orbited the Sun instead of the Earth. Lacking a significant atmosphere and being largely devoid of water and organics, the Moon is not itself considered to be a viable host for indigenous life. But the presence of such a massive near-neighbor clearly had, and continues to have, an influence on the Earth and its evolution, so it is important to consider whether the Moon has any direct relevance to the Earth's habitability.

doi:10.1088/978-1-6817-4676-0ch5

To the best of our knowledge, the first 100 Ma or so of the Earth's history were dominated by two major events: core formation and Moon formation (see figure 1.1). The timings of these events are not known precisely, but it is probable that core formation—the gravitational separation of heavier elements, especially iron, that sink toward the center—was already well advanced before the birth of the Moon (section 5.3.1). Gravitational capture by the Earth of a preformed Moon can be ruled out as dynamically implausible and inconsistent with what we have learned of the Moon's composition, notably its low iron content. The proposal that best explains all known properties, including the lunar composition and the relatively high angular momentum of the system, is the giant impact hypothesis (see Canup 2004 for a review). It is proposed that the Earth was struck a glancing blow by another planet (Theia), similar in mass to present-day Mars. (We may speculate that if the collision had been head-on the Earth might have been destroyed.) Assuming that Theia and the Earth were each already differentiated, at least partially, into an iron-rich core and a silicon-rich mantle, the model provides a natural explanation for the relatively low density and low iron content of the Moon. The force of the impact in supposed to have ripped mantle material from both Theia and (to perhaps a lesser extent) the Earth, generating a circumplanetary disk that subsequently condensed to become the Moon. Theia's core is thought to have coalesced with the Earth a few hours after the initial impact.

The oblique angle of the impact boosted the Earth's rate of rotation, resulting in a spin period that may have been as short as ~5 h in the immediate aftermath of this event. Over the eons since then, tidal forces have resulted in a gradual transfer of angular momentum from the Earth to the Moon. One outcome of this is a slowing of the Earth's rotation rate, thereby increasing the length of the day. Precise estimates of the rate of increase are not possible for early epochs, but if it was similar to estimates for more recent times, an approximately 10 hour-day is suggested for the early Archean eon ~1 Ga after Moon-formation. This loss of angular momentum by the Earth is balanced by an acceleration of the Moon's orbital speed, causing it to slowly spiral outward. Tides were also responsible for slowing the Moon's rotation rate, until it stabilized in its current state of 'locked rotation' with the spin period equal to the orbital period, causing the same hemisphere always to face toward the Earth.

Assuming the model is broadly correct, its predictions have evident consequences for surface conditions on the early Earth. Initially, the entire surface would have been an ocean of molten magma. Much of the primordial atmosphere may have been lost, and any pre-existing surface water would have been vaporized. Oceans accumulating subsequent to crust formation would have been subject to enormous tides, ebbing and flowing at more than double the present-day frequency. Whether such tides were helpful or detrimental to prebiotic evolution in surface water is uncertain: some turbulence is helpful, but stability is also needed, and this would surely have been lacking in coastal waters at early times. In contrast, hydrothermal systems on the ocean floor are effectively immune to this problem, generating dynamic environments with a degree of stability unaffected by surface tides.

Another potentially relevant influence of the Moon is its effect on the Earth's obliquity, i.e. the axial tilt of its spin axis. The Earth's spin axis is currently tilted at an angle of 23.4° relative to the perpendicular to its orbital plane, and this configuration is responsible for our seasons. The angle changes over time, oscillating between 22.1° and 24.5° with a period of about 41 thousand years, an insufficient variation to have any major consequences for global climate. But if it were to oscillate with much greater amplitude (as in the case of Mars), or to stabilize at a far less favorable orientation (such as the ~82° obliquity of Uranus), this would surely be detrimental and possibly disastrous for life. Long-term variations in the obliquity of a planet arise from the net effect of its gravitational interactions with all other members of the system over time, and may be amplified by factors such as orbital resonances, chance alignments, close encounters and collisions. The presence of a massive moon tends to be a stabilizing influence, minimizing the amplitude of such variations, and because of this it has sometimes been argued that the presence of the Moon is essential for life. However, calculations by Lissauer (2012) and colleagues suggest that the obliquity variations of a moonless Earth would be insufficient to pose a major threat to habitability.

In summary, there does not seem to be a compelling reason to doubt that life could have formed and thrived on Earth if it had no Moon. The Moon is nevertheless an object of considerable interest to astrobiologists, preserving a record of past events presumed to have affected both members of the Earth–Moon system, as discussed in the next section.

5.2 Cosmic impacts

5.2.1 Significance

The Earth is continually bombarded by interplanetary debris. Most of these particles are very small: the frequency varies approximately in proportion to D^{-2} for particles of diameter D. Their fate on entering the atmosphere is dependent on size. The smallest (1–100 μm) dust grains decelerate rapidly in the stratosphere and permeate gently to the surface; somewhat larger (0.1–10 mm) particles are completely vaporized by atmospheric friction, still larger ones may penetrate the atmosphere to strike the surface. Depending on their size, those in the latter group may deliver anything from a small meteorite to environmental catastrophe.

Clearly, in the astrobiological context, these events have the potential to be either beneficial or detrimental. Non-destructive accumulation of asteroidal and cometary debris may well have been a significant and perhaps vital means of replenishing the planet's endowment of water and organic molecules subsequent to the Moon-forming event; but collisions with larger projectiles pose an evident threat. Those with diameters $\gtrsim 1$ km carry enough momentum to rip through the atmosphere with very little deceleration, hitting the surface at almost their cosmic speeds (10–80 km s^{-1}). The kinetic energy released by such an impact is analogous to the detonation of a bomb of equivalent energy, resulting in the excavation of a crater a factor of 10 or more greater in size than the diameter of the projectile. The consequences for the environment are by no means restricted to the area of impact: large impacts may generate global tsunami, seismic shocks, wildfires and dust storms, resulting in climate change on the scale predicted for a 'nuclear winter' scenario. A well-known example is the Cretaceous–Tertiary event implicated in the extinction of the dinosaurs 65 Ma ago, attributed to the impact of a comet or asteroid approximately 10 km in size. Even larger bodies may deliver enough energy to vaporize the oceans.

An assessment of the importance of such events for the origin and survival of life requires an understanding of the impact history of the Earth. Impacts are in essence the final phase of the accretion process that formed the planets (section 4.2.3), and the rate would have declined rapidly as the protoplanetary disk was cleared out. Nevertheless, impacts continue at a low level to this day, the result of the dynamic nature of the solar system, which can, for example, cause asteroids from the main belt to be deflected into Earth-crossing orbits. Some 200 impact craters have been identified on Earth, with estimated ages $\lesssim 500$ Ma. There can be no doubt that the Earth was subject to bombardment at much earlier times, but that record has been erased by erosion, volcanism, and subduction of continental crust. Lacking an atmosphere, hydrosphere and subduction zones, and with little or no volcanic activity in the last 3 Ga (see figure 4.8), the Moon retains information that has been eradicated on Earth. To gain insight into the impact history of the Earth we must use its nearest neighbor as a proxy.

5.2.2 Cratering chronology

Because impact craters may be erased by the endogenous activity of a body, the distribution of craters provides a measure of the relative age of a surface. An example is shown in figure 5.1 for an area of the Moon that includes a typical example of a lunar volcanic plain (Mare Nectaris). The entire area of the image was presumably subject to intense cratering at early times (see figure 5.2), but some have since been erased. Mare Nectaris is thought to be a basin excavated by a giant

Figure 5.1. Comparison of cratering in Mare Nectaris and adjacent highland regions of the Moon. Mare Nectaris (the approximately circular, dark area to the right of center) is about 340 km in diameter; it contains relatively few impact craters because it was resurfaced by volcanism some 3.5 Ga ago. The more densely cratered highlands to the left contain a record of impacts from earlier times. Image credit: NASA/JPL (Lunar Reconnaissance Orbiter).

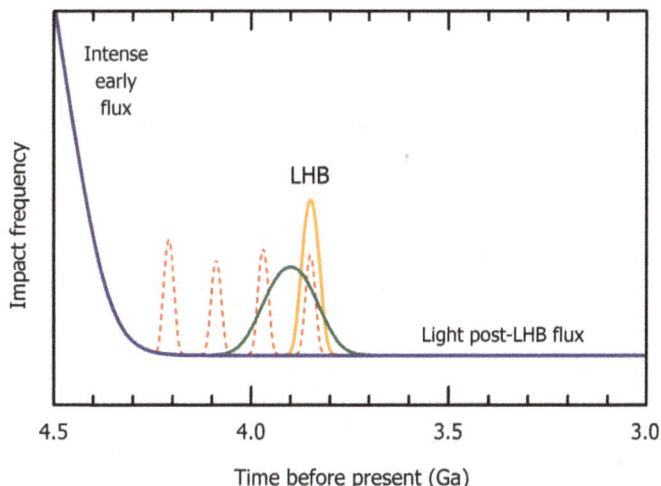

Figure 5.2. Plot of impact frequency versus time, illustrating some possible interpretations of the lunar cratering record. The blue curve represents a smooth, steep decline from intense early bombardment to the lower level implied by light cratering of the lunar maria (figure 5.1). The other curves suggest a range of models to explain the late heavy bombardment (LHB) ~3.85 Ga ago, including a narrow single peak (orange), a broader peak (green), and a sequence of sporadic events (dotted red).

impact that occurred about 3.9 Ga ago. Subsequently, during the next few hundred million years, volcanism led to the basin filling with magma to form a roughly circular volcanic plain[1]. Scrutiny of the image shows that craters are rather few (and mostly very small) within the area of the plain itself: it displays a record only of impacts that occurred subsequent to the episode of volcanism that formed the plain. In contrast, the highland area to the left contains a much larger number of craters and a wider range of sizes, as it preserves a record of the impact flux over a longer time span, dating back to earlier times. A quantitative measure of such differences may be determined by counting the number of craters per unit area. This is a useful technique for study not only of the Moon but more generally of cratered bodies throughout the solar system, including Mars (chapter 7), and icy worlds that are subject to resurfacing by cryovolcanic activity (chapter 8).

It is important to remember that ages obtained by this technique are relative, not absolute: it enables younger surfaces to be distinguish from older ones without direct measurement of their actual ages. However, we do have samples from certain areas of the Moon that can be analyzed in the laboratory to determine absolute ages by the radiometric technique (section 1.3). Results show that rocks from the lunar highlands are generally >4.0 Ga old, whereas basalts from the volcanic plains are significantly younger, with ages typically in the range 3.2–3.8 Ga.

Radiometric analyses of basalts from the lunar maria measure the time elapsed since solidification of the magma that filled the basins; but of far greater interest in

---

[1] Lunar volcanism was dominated by fluid, basaltic magma that spread over large distances rather than building volcanic mountains.

the astrobiological context is the timing of the impacts that formed the basins prior to volcanism. To address the latter question, investigators attempt to identify samples with radiometric clocks that were reset by melting driven by the heat of impact rather than by volcanism. These studies focus on samples of lunar regolith (loose surface material) and breccias (rocks composed of compressed mineral fragments in a grainy matrix). Technical advances in the intervening years have greatly enhanced the information that can be extracted from such samples returned by the Apollo and Luna missions more than 40 years ago.

Contained within the samples are small (~200 μm), glassy spherules with compositions consistent with an origin in impact-melted surface material rather than volcanic emissions. In principle, a picture of lunar impact frequency over time can be built up by determining radiometric ages for a large number of these spherules collected from various areas of the Moon. In practice, however, there are potential problems with both the analysis and the interpretation of results (the latter discussed in section 5.2.3 below). One of the primary radiogenic isotopes used in the analysis, ^{40}Ar, is produced by decay of ^{40}K with a half-life of 1.25 Ga. Argon is, of course, a noble gas: ideally, the radiogenic ^{40}Ar produced within each spherule should remain trapped inside prior to analysis, but in practice there is likely to be some loss over time caused by diffusion, the rate of which depends on temperature. If uncorrected, this loss would lead to a systematic error such that the apparent age underestimates the true age. Corrections based on laboratory measurements of diffusion rates can be applied, but with added uncertainty. Some possible interpretations of the available data are shown in figure 5.2.

5.2.3 Late heavy bombardment

Results from many studies, conducted over the years since the Apollo and Luna samples became available, have suggested that a peak in the lunar cratering rate occurred over a relatively short time period approximately 3.8–3.9 Ga ago (see figure 5.2). The estimated timings of several basin-forming impacts appear to fall within this period, including those responsible for the Mare Crisium, Imbrium, Nectaris, Orientale and Serenitatis. Occurring more than 600 Ma after the birth of the solar system, these impacts cannot be attributed to a late phase of normal planet formation in the solar nebula, as the protoplanetary disk would have dissipated long before then. Interpretation has instead focused on a hypothetical event that caused a major disturbance in the comet, asteroid or Kuiper belt populations at this time, scattering many of them into orbits that led to collisions with the Moon and planets, a phenomenon termed the late heavy bombardment.

The probable cause of such a disturbance is migration of the giant planets from their initial orbits, as the result of their gravitational interactions with each other and with other bodies. Once the protoplanetary disk had dissipated, such migrations would generally have been small, but subject to amplification by orbital resonances. A resonance occurs when the orbital period of one body is a simple multiple of another, leading to perturbations that are systematic rather than random. In a hypothesis called the Nice model[2], it is proposed that a major disruption occurred when a 1:2 resonance became established between the orbits of Jupiter and Saturn. This led to Jupiter and Saturn migrating inward and outward, respectively (the ratio of their orbital periods is now 2.48); Uranus and Neptune also migrated outward (they may even have switched places according to some simulations), and this caused major disruption of the Kuiper-belt objects. Another likely outcome is disturbance of the main asteroid belt by the inward migration of Jupiter, with asteroids that fell into an orbital resonance with Jupiter most affected. The asteroid belt now contains annular 'zones of avoidance' (Kirkwood gaps) that correspond to resonances with Jupiter's current orbit.

It should be noted, however, that interpretation of the lunar data as an isolated, narrow peak in the impact flux ~3.85 Ga ago has been challenged. A potential source of bias arises from the discovery that one of the major impacts (the Imbrium event, dated at 3.85 Ga) was particularly pervasive in scattering ejecta to other regions of the Moon from which samples have been collected and analyzed. Nevertheless, the evidence is clear that the Moon suffered a series of major impacts that persisted long after its birth, perhaps in a sporadic manner rather than in a cluster, up until about 3.8 Ga ago. Examples of some possible models are illustrated in figure 5.2. It seems inevitable that the Earth suffered a similar fate, and this is likely to have been a major challenge for the early emergence and survival of life.

5.3 Emergence of the atmosphere and hydrosphere

Models predict that, following the Moon-forming event, the Earth's surface was a completely molten magma ocean at a temperature ~3000–5000 K, surrounded by a hot atmosphere composed largely of vaporized silicate rock. Over the next few million years, as the planet cooled, silicates precipitated out of the atmosphere and the magma began to form a veneer of solid crust. The atmosphere was replenished by copious release of volcanic gases, resulting in pressures that may have been 10–100 times greater than present-day values. As the crust solidified, surface water began to accumulate. The extent to which the Earth retained its pre-existing water through the Moon-forming event is uncertain: at such high temperatures, H_2O would have been at least partially dissociated, with subsequent loss of H to space; but some may have survived or recombined as steam in the hot early atmosphere, to precipitate during the cooling phase. Other likely contributors to the growth of the hydrosphere include water from ongoing volcanic activity and infalling debris from space. The oldest available samples of the Earth's crust (zircon crystals) show evidence of having formed in aqueous environments, confirming that surface water was already present in the early Hadean eon, about 4.4 Ga ago.

The composition of the early atmosphere is a vital factor that governs the efficiency of organic synthesis in surface environments on the prebiotic Earth, as discussed in section 1.4.1. The key question is whether the atmosphere was broadly reducing (rich in hydrogenated molecules such as CH_4 and NH_3) or non-reducing (in

---

[2] Named after the French city, pronounced 'Niece'.

which CO_2 and N_2 would have been the dominant carriers of C and N). The Miller–Urey experiment (figure 1.2) demonstrated that a reducing atmosphere is conducive to abiotic synthesis of prebiotic molecules, but a non-reducing atmosphere is not. The oldest samples of the Earth's crust carry a chemical memory that allows us to address this question.

5.3.1 The message in the rocks

Gases trapped within the magma of an erupting volcano are released to the atmosphere as the result of decompression and reduced solubility as the rising magma cools. The composition of the gases within the magma is closely linked to the composition of the magma itself, specifically its oxidation (redox) state. An iron-rich mantle in a low oxidation state may generate free hydrogen by abstracting oxygen from water:

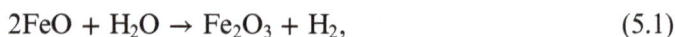

$$2FeO + H_2O \rightarrow Fe_2O_3 + H_2, \tag{5.1}$$

where the iron has transitioned from the ferrous (valence +2) to the ferric (valence +3) state. The resultant free hydrogen may participate in further reactions prior to eruption, generating hydrogenated gases such as CH_4, NH_3 and H_2S. But if the mantle is depleted in Fe as the result of core formation (figure 5.3), as it is today, the Fe that remains in the mantle is typically in a more highly oxidized state (Fe_3O_4 and Fe_2O_3 rather than of FeO), rendering it far less effective as a crucible for reducing water to hydrogen. If follows that the timing of core formation is an important factor in determining the redox state of the Earth's early atmosphere.

As discussed in section 5.1 above, the core formation process is presumed to have been already completed or well advanced by the time of the Moon-forming event 4.5 Ga ago, in order to explain the low density and low iron content of the Moon. Isotopic analyses of terrestrial igneous rocks enable this assumption to be tested. The radiogenic tungsten isotope ^{182}W is formed by decay of ^{182}Hf (hafnium) with a half-life of 9 Ma. Hf and W are expected to segregate during core formation: Hf has an affinity for O and remains in the mantle, whereas W has an affinity for Fe and sinks to the core. Because of this, the abundance of ^{182}W relative to other (non-radiogenic)

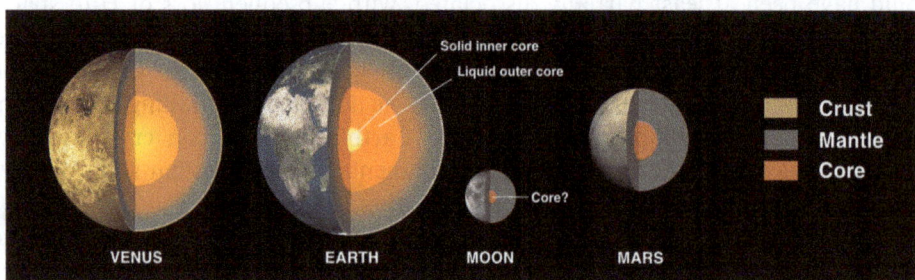

Figure 5.3. Models for the internal structure of Venus, Earth, Moon and Mars. Formation of the Earth's dense, iron-rich core is thought to have been well advanced prior to the Moon-forming impact, consistent with the relatively low density (and small core size) of the Moon. The Earth's magnetic field results from convection currents in the liquid outer core, lacking in the other bodies. Image credit: NASA/Wikimedia Commons.

tungsten isotopes is essentially a chronometer for core formation: results show that significant ^{182}W was generated after core formation but before ^{182}Hf became extinct (as it effectively would have been after an interval of a few half-lives). This suggests that the process was completed on a timescale of about 30 Ma. The rapidity of core formation implied by these results gives us cause to suspect that the early atmosphere was non-reducing. Independent confirmation is sought by geochemical analyses that provide direct evidence of magma redox state at earlier times.

Modern terrestrial volcanoes fall into two broad categories: (i) shield volcanoes associated with isolated 'hot spots' or divergent tectonic plate boundaries, in which basaltic magma originates from deep in the mantle (figure 5.4); and (ii) conical (strato) volcanoes, mostly associated with convergent plate boundaries, in which the magma is largely reprocessed crust (Mauna Loa and Mount Etna are classic examples of each type, respectively). Of these, shield volcanoes represent a more general form of volcanism known to exist on other bodies in the solar system, including Venus, Mars, and Jupiter's moon Io, whereas strato volcanoes are known to exist only on Earth and in regions near tectonic plate boundaries. Volcanic gases released during the Hadean eon presumably originated from the mantle, so shield volcanoes are considered appropriate modern analogs of volcanism on the early Earth. Emissions from present-day shield volcanoes (e.g. figure 5.5) are dominated by H_2O ($\sim$60%), CO_2 ($\sim$25%), and SO_2 ($\sim$10%), where the abundances are very approximate averages by volume; other gases present in lesser amounts include N_2, Ar, CO, H_2S, HCl and H_2. This mean composition is consistent with the high oxidation state of recently erupted magma, which serves as a benchmark for comparison with magmas erupted at much earlier times.

Geochemical methods have been developed that enable the redox state of ancient igneous rock samples to be quantified. The ideal would be to directly measure the ratio Fe^{2+}/Fe^{3+} (where the superscripts denote valency) in samples of known age,

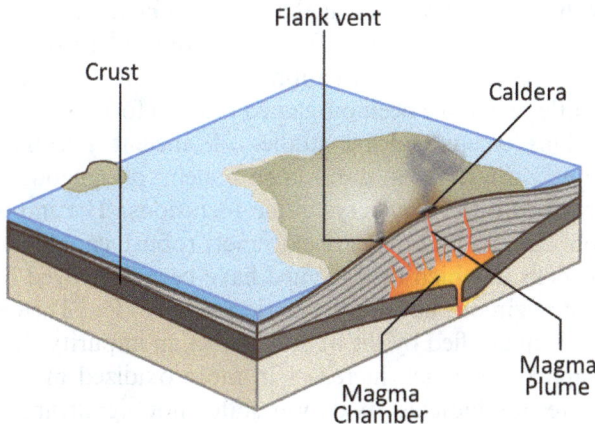

Figure 5.4. Schematic view of a terrestrial oceanic shield volcano. The magma chamber is fed by mantle material rising through an opening in the crust. Unlike other types of terrestrial volcano, shield volcanoes may form in regions remote from tectonic plate boundaries and have counterparts on other bodies in the solar system. Image credit: Niamh O'C/Resident Mario (Creative Commons).

Figure 5.5. A view inside the summit caldera of Nyamuragira, an active shield volcano associated with the East African rift system (dated April 2, 2015). The volcanic gases released to the atmosphere are predominately CO_2, SO_2 and water vapor. Image credit: MONUSCO/Abel Kavanagh (Wikimedia Commons).

but the oxidation state of Fe is liable to alteration subsequent to eruption. However, other metallic elements such as Cr provide reliable proxies for Fe. Rocks from the Acasta region of northwestern Canada and the Akilia Island, Greenland, dated at 3.96 and 3.85 Ga, respectively, have been studied (see Delano 2001). These and other samples with radiometric ages in the 3.5–3.9 Ga range have been found to have oxidation states indistinguishable from those of modern basalts.

It is important to extend this investigation to earlier times, as much of the volatile content of the mantle may have been outgassed prior to formation of the Acasta and Akilia samples. This is enabled by studies of ancient zircon crystals. Zircon (zirconium silicate, $ZrSiO_4$) is a minor component of igneous rocks, typically condensing as small (0.1–0.3 mm) crystalline inclusions. The importance of these crystals stems from the fact that they are extremely robust, capable of surviving long after the igneous rocks in which they formed have been destroyed by heat, pressure or erosion. The oldest zircons have radiometric ages ~4.4 Ga. The oxidation state of the parent magma is quantified by the abundance of an impurity element, cerium, in the zircons, the uptake of which increases in more oxidized melts (see Trail *et al* 2011). Again the results indicate oxidation states not significantly different from present-day values.

In summary, the Earth is thought to have acquired a dense early atmosphere composed largely of CO_2 and water vapor. This atmosphere would have resulted in a strong greenhouse effect, thus ensuring that global surface temperatures warm enough for H_2O to exist in liquid form were maintained at a time when the Sun's luminosity was 30% less than today (section 4.4.1). Atmospheric pressure would have been moderated by precipitation of H_2O into the early oceans, and by loss of CO_2 arising from its solubility in water and geochemical processing into carbonate minerals. Molecular nitrogen, although no more than a minor component of present-day volcanic emissions, is presumed to have accumulated steadily over time, persisting in the atmosphere as the concentrations of H_2O and CO_2 declined, until it became the dominant constituent that it is today. The volcanic emissions are notably devoid of free oxygen, as the result of its propensity to oxidize other elements present in the magma. Dissociation of H_2O in the upper atmosphere by solar radiation may have produced small amounts at early times, but the rise of atmospheric O_2 could not begin until photosynthetic life had formed.

5.3.2 Origin of the oceans

The Earth's lakes, seas and oceans contribute about 0.02% of the mass of the planet. A substantial but poorly constrained additional mass of H_2O exists within the body of the Earth, perhaps comparable with or exceeding that on the surface. In addition to covering ~70% of the surface, this 'minor' ingredient (in terms of mass) exerts a major influence on the functionality of the planet, serving as a lubricant for tectonic plate movement, a regulator of temperature and climate, an agent for erosion and sedimentation, a medium for chemical and biochemical reactions, and an environment for life. Compared with its nearest planetary neighbors, the Earth is richly endowed with water: if Venus and Mars received similar initial endowments they have been mostly squandered. Photodissociation by solar UV radiation and leakage into space are probable causes, accelerated by the solar wind and, in the case of Mars, by a relatively weak gravitational field (see figure 4.4). Unlike its neighbors, the Earth also possesses a magnetic field, generated in its liquid outer core (figure 5.3), that is strong enough to shield it from the harmful effects of energetic ions in the solar wind, and this may have contributed to its ability to retain its H_2O.

But what was the source of this water? This apparently simple question is surprisingly difficult to answer. Was it mostly contained within the original aggregate that formed the Earth's bulk, or was it mostly acquired later, over a period extending to times long after the Moon-forming event? With a feeding zone well within the frost line of the solar nebula (section 4.2.2), the planetesimals that formed the original aggregate would have been devoid of ice and hydrated minerals. This being the case, let us consider two remaining possibilities, summarized below. It should be noted that these proposals are by no means mutually exclusive:

- Gaseous OH or H_2O molecules in the solar nebula were absorbed into the Earth as it formed. The Earth would not have gained enough mass to allow the possibility of direct capture from the nebular gas until the formation process was almost complete, by which time the gaseous disk would have largely dispersed. A more likely possibility is that dust grains in the nebula were carriers, acquiring stable surface coatings of H_2O or (more probably) OH prior to accretion into the Earth's birth aggregate. An abundance of OH

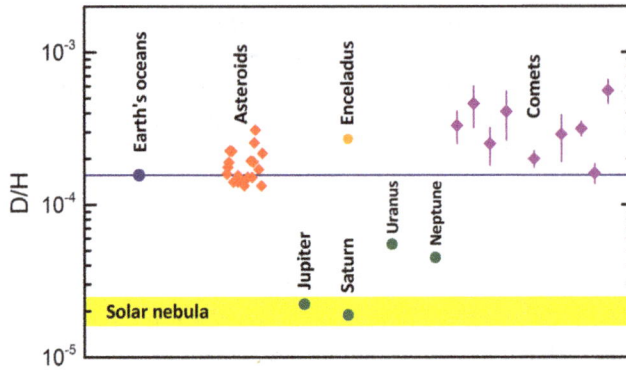

Figure 5.6. Deuterium abundances in the solar system, comparing the mean value for the Earth's ocean water (blue circle and horizontal blue line) with H_2O reservoirs in asteroidal hydrated silicates, cometary ices, and Saturn's moon Enceladus (D/H values indicate the number of HDO molecules relative to H_2O). Data for molecular hydrogen in the atmospheres of the giant planets are also shown. Data are displayed in order of increasing solar formation distance from left to right (not well constrained for comets). The horizontal yellow band indicates the D/H ratio for unfractionated solar nebular gas. Based on data from Altwegg *et al* (2015) and references therein.

in the Earth's interior would naturally lead to production of H_2O (2 OH $\rightarrow$ H_2O + O), which would feed into volcanic gas emissions.

- The Earth acquired volatile-rich solids that originated at greater solar distances. Potential sources include comets (section 4.3.1), and volatile-rich asteroids (section 4.3.2), which contain H_2O primarily in the form of ice and mineral hydration, respectively. Delivery depends on perturbations and collisions that carry these bodies into the inner solar system, the rate of which is influenced by the timing of formation and subsequent migrations of the giant planets (sections 4.2.4 and 5.2.3). Comets are richer in H_2O than asteroids, but the orbital speeds of asteroids relative to the Earth tend to be less extreme, increasing the probability of constructive capture.

An important clue is provided by the deuterium abundance of the Earth's oceans in comparison to these potential sources, illustrated in figure 5.6. The baseline indicated by the yellow band represents unfractionated solar nebular gas inherited from the parent cloud[3]. Data for asteroidal hydrated silicates (deduced from analyses of meteorites) and cometary ices are included. As discussed in section 3.1.4, deuterium fractionation resulting in enhanced D/H is a signature of interstellar chemistry at low temperature. H_2O molecules inherited from the parent cloud may retain this signature in the cold outer regions of the solar nebula, where the ice remains frozen; but inside the frost line, where the ice is sublimated and

---

[3] D/H values for the giant planets Jupiter and Saturn are consistent with the prediction that most of their mass accumulated from unfractionated nebular gas (section 4.2.4), whereas Uranus and Neptune acquired somewhat higher proportions of fractionated ices. The value for Enceladus is consistent with cometary values, suggesting similar source material.

mixed with hot hydrogen gas, the signature is largely erased by isotopic exchange reactions such as

$$HDO + H_2 \rightarrow H_2O + HD. \tag{5.2}$$

Because of this, D/H levels in gas-phase OH and H_2O available in the Earth's formation zone were probably close to the baseline. The Earth's oceans are enhanced relative to this level by a factor of about eight. On the assumption that there has been no major change in D/H since the oceans formed, this result appears to exclude nebular gas as the dominant source of the Earth's water.

In contrast, cometary D/H values are generally greater than the terrestrial value (the mean is approximately double). Taken in isolation, this result suggests that comets may have contributed but are unlikely to have been the only source of the Earth's water. A further and more rigorous constraint is provided by consideration of noble gases such as argon: available measurements suggest that Ar is present in comets at approximately the solar abundance level (figure 2.1), much higher than on Earth. If comets were a major source of terrestrial water we should expect to find far more argon in the present-day atmosphere.

So we are left with hydrated asteroids, which do indeed provide a close match to the terrestrial D/H (figure 5.6). However, again, other evidence leads us to doubt that asteroids are the sole solution to the problem. Because of their modest water content ($\lesssim 15\%$), accumulation of a very large mass of asteroids would be needed to deliver enough H_2O to account for the present-day oceans and subsurface H_2O. If this was so, a considerably larger mass would also have been delivered in the form of minerals. However, comparisons of isotopic abundances for elements such as oxygen and osmium in asteroidal and terrestrial minerals indicate a better match with anhydrous classes of stony meteorites ($\lesssim 0.1\%$ H_2O) than with the water-bearing carbonaceous chondrites.

Taking all of these caveats into consideration, it appears that no single source provides a satisfactory explanation for the Earth's water. A compound model that includes contributions from different sources seems more promising. A mix dominated by asteroids (both hydrous and anhydrous) with smaller contributions from comets and nebular gas, in proportions that yield the terrestrial mean D/H ratio, may avoid violating other constraints. See Izidoro *et al* (2013) for examples.

5.4 Toward a prebiotic world

A picture is emerging of conditions on the Earth in the Hadean eon, from 4.5 to 4.0 billion years ago. The first continental crust and oceans appeared early in this time-frame, and a dense CO_2-rich atmosphere accumulated from volcanic gases, its greenhouse effect compensating for the lower solar luminosity to yield reasonably clement temperatures. The greater amplitude and frequency of lunar tides compared with today caused turbulent motion and mixing of the larger bodies of surface water. The water was presumably acidic, as dissolved CO_2 reacts with H_2O to form carbonic acid (H_2CO_3), and atmospheric SO_2 (a probable component of the volcanic emissions) may be processed to sulfuric acid (H_2SO_4), resulting in acid rain.

Meanwhile, asteroidal and cometary debris continued to arrive from space, probably at much higher rates than today, and the threat of planet-wide disruption arising from occasional large impacts persisted throughout this period.

The next step toward life is the emergence of a prebiotic world (section 1.4.1), in which the basic molecular ingredients needed for life are present in environments conducive to their polymerization. If our deduction that the early atmosphere was non-reducing is broadly correct, atmospheric processes seem unlikely to have been major contributors to the inventory of prebiotic molecules on the early Earth. This conclusion holds regardless of the energy source assumed to drive chemical reactions (e.g. solar radiation, electric discharge, impact shocks). In this section we assess the main alternatives: exogenous delivery, and endogenous production by geochemical processes.

5.4.1 Organics from space

The Earth is continuously accreting carbon and organic molecules from interplanetary space, and this has presumably been the case throughout history. Projectiles capable of delivering them intact to the surface (section 5.2.1) take two distinct forms: dust grains, probably mostly of cometary origin (section 4.3.1), and carbonaceous meteorites (section 4.3.2). Volatiles are unlikely to survive impacts at speeds above about 10 km s^{-1}, and this excludes delivery by bodies greater than about 1 km in diameter if we assume present-day atmospheric density and pressure. The modern accretion rate of unablated carbonaceous matter by all particles in the appropriate size range is estimated (Anders 1989) to be approximately $3 \times 10^5 \text{ kg yr}^{-1}$. If this rate remained constant over the entire lifespan of the planet, a total of $1.4 \times 10^{15} \text{ kg}$ would have been delivered to date, roughly comparable with the Earth's current biomass[4]. The mean accretion rate in the Hadean is unknown but was presumably much higher.

Results discussed in section 5.3.2 above indicate that volatile-rich asteroids were an important and perhaps dominant source of the Earth's oceans. This being the case, a simple calculation allows the amount of organic carbon that accompanied the water to be estimated. Assuming that carbonaceous chondrites are representative of the composition of these asteroids, the average mass ratio of H_2O to organic matter (in all forms) is about 4:1. If the asteroids delivered a total mass of H_2O equal to the mass of the present-day oceans ($\sim 1.5 \times 10^{21} \text{ kg}$), the accompanying mass of organic matter would amount to $3.8 \times 10^{20} \text{ kg}$, assuming that organics are as likely as H_2O to survive atmospheric entry and impact. Of this total, perhaps some 10%–20% is in forms potentially most useful to prebiotic chemistry, such as amino acids, carboxylic acids, purines and sugar-related compounds, the rest being mostly insoluble organic refractory matter (see section 4.3.2). Nevertheless, this is an enormous mass of prebiotic molecules, exceeding the mass of the current biosphere by several orders of magnitude. This conclusion would still hold if comets made a

---

[4] The mass of all organic matter in the Earth's current biosphere, estimated to be $\sim 2 \times 10^{15} \text{ kg}$; see Bodnar *et al* (2013) and references therein.

larger contribution to the oceans, as their organic fraction relative to H_2O is probably similar to within a factor ~2.

In summary, the influx of prebiotic organic matter to the early Earth from interplanetary sources was potentially extremely large. Delivered by myriads of small particles in preference to sparser large ones, it would initially have been distributed rather evenly over the entire surface of the planet. Subsequent redistribution, presumably driven by flow and evaporation/precipitation cycles in surface water, may have created local concentrations conducive to chemical evolution. Under conditions of high temperature and pressure, even the organic refractory component may yield useful reactants such as NH_3, as demonstrated in laboratory simulations of hydrothermal environments by Pizzarello and Williams (2012).

5.4.2 Geochemical processes

Under suitable conditions, organic molecules may readily be synthesized by interactions between certain minerals and an inorganic form of carbon (typically CO_2) in the presence of water. Possibilities include catalytic surface reactions, exothermic processes such as serpentinization (see below), and endothermic processes driven by geothermal heat. Likely geological settings include hydrothermal systems, in which surface or subsurface water is in contact with or close proximity to a geothermal energy source. Oceanic hydrothermal systems, in particular, have received much attention as promising sites for endogenous production of molecules essential to life and, indeed, for life itself (see chapter 6 for a detailed discussion).

Serpentinization is a metamorphic process in which Fe- and Mg-rich silicates (olivines and pyroxenes) are oxidized by abstraction of O from water, generating free H or hydrogenated carbon, with a net release of energy (see Holm *et al* 2015 for a review). Although often discussed in the context of hydrothermal systems, as an exothermic process serpentinization can occur over a wide range of physical conditions. To give examples, oxidation of fayalite to magnetite and silica releases H_2:

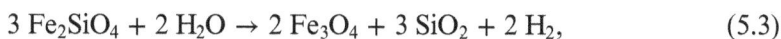

$$3\,Fe_2SiO_4 + 2\,H_2O \rightarrow 2\,Fe_3O_4 + 3\,SiO_2 + 2\,H_2, \qquad (5.3)$$

and oxidation of olivine to magnetite and serpentinite in the presence of CO_2 releases CH_4 by means of a process that may be summarized:

$$(Fe,\,Mg)_2SiO_4 + nH_2O + CO_2 \rightarrow Fe_3O_4 + Mg_3Si_2O_5(OH)_4 + CH_4. \qquad (5.4)$$

Magnesium carbonate ($MgCO_3$) may also be produced by these reactants, the outcome depending on the Mg:Fe ratio of the olivine. Silicates capable of undergoing serpentinization are very abundant in the Earth's crust, and it is feasible to suppose that the process has been active ever since the continents and oceans formed. The products initiate new pathways for synthesis of organic molecules.

Catalytic reactions on mineral or metal surfaces may accelerate organic synthesis in geothermal environments. An important example is the Fischer–Tropsch process, a set of reactions that combines H_2 with CO (or CO_2) to produce hydrocarbons. CO may be formed by reduction of CO_2:

$$CO_2 + H_2 \rightarrow CO + H_2O, \tag{5.5}$$

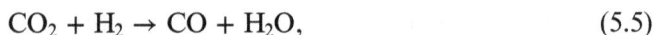

and CO may be processed to hydrocarbons by the generic reaction set:

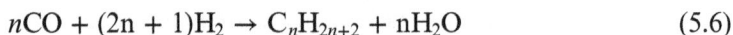

$$nCO + (2n + 1)H_2 \rightarrow C_nH_{2n+2} + nH_2O \tag{5.6}$$

where n is a positive integer ($n = 1$ yields methane, $n = 2$ yields ethane, etc). The Fischer–Tropsch process is particularly significant as an abiotic means of forming the linear fatty acids present in cell walls. Amino acids and other N-bearing prebiotic molecules may also be produced if NH_3 is available as a reactant. Such processes may well have contributed to the production of prebiotic molecules on other bodies in the solar system, such as Mars, icy worlds and the parent bodies of the carbonaceous chondrites, as well as on the Earth.

Questions and discussion topics

- Why is an impact crater much larger than the projectile that formed it?
- Consider the effect of trajectory for projectiles hitting the Earth by comparing the probable outcome for a very large bolide (100 km) and a relatively small one (1 km) in head-on and oblique collisions.
- List all the reasons you can think of for a large asteroid colliding with the Earth to cause mass extinctions. Would you expect the threat to be equally grave for all species on Earth, regardless of their form, size and habitat, or are some species much more vulnerable than others?
- Consider what the consequences might be for life on Earth if the Earth's obliquity were similar to that of Uranus.
- Why is it important to know when the Earth's core formed?
- Consider how you might attempt to estimate the total biomass of the Earth.
- Compare the probable distributions of organic molecules on the early Earth from exogenous and endogenous sources. Consider how local concentrations in aqueous environments conducive to prebiotic evolution might arise in each case.

References and further reading

Altwegg K *et al* 2015 67P/Churyumov-Gerasimenko, a Jupiter family comet with a high D/H ratio *Science* **347** 1261952

Anders E 1989 Pre-biotic organic matter from comets and asteroids *Nature* **342** 255

Bodnar R J *et al* 2013 Whole Earth geohydrologic cycle, from the clouds to the core: The distribution of water in the dynamic Earth system *Geol. Soc. Am. Spec. Pap.* **500** 431

Canup R 2004 Origin of terrestrial planets and the Earth-Moon system *Phys. Today* **57** 56

Colin-Garcia M, Heredia A and Cordero G 2016 Hydrothermal vents and prebiotic chemistry: A review *Bull. Mex. Geolog. Soc* **68** 599

Delano J W 2001 Redox history of the Earthas interior since 3900 Ma: Implications for prebiotic molecules *Orig. Life Evol. Biosph.* **31** 311

Drake M J and Righter K 2004 Determining the composition of the Earth *Nature* **416** 39

Holm N G *et al* 2015 Serpentinization and the formation of H_2 and CH_4 on celestial bodies (planets, moons, comets) *Astrobiology* **15** 587

Izidoro A, de Souza Torres K, Winter O C and Haghighipour N 2013 A compound model for the origin of Earth's water *Astrophys. J.* **767** 54

Lissauer J J, Barnes J W and Chambers J E 2012 Obliquity variations of a moonless Earth *Icarus* **217** 77

Pizzarello S and Williams L B 2012 Ammonia in the early Solar System: An account from carbonaceous meteorites *Astrophys. J.* **749** 161

Trail D, Watson E B and Tailby N D 2011 The oxidation state of Hadean magmas and implications for early Earthas atmosphere *Nature* **480** 79

Whittet D C B 1997 Is extraterrestrial organic matter relevant to the origin of life on Earth? *Orig. Life Evol. Biosph.* **27** 249

IOP Concise Physics

Origins of Life
A cosmic perspective
Douglas Whittet

Chapter 6

The origin of terrestrial life: converging on a paradigm

Research described in the previous chapter provides insight into conditions prevailing in the earliest epoch of Earth's history. Unlike the hellish place suggested by its name, the Hadean Earth appears to have become reasonably temperate within about 100–200 million years of its birth, by which time the first continents and oceans had formed, and a moist, CO_2-rich atmosphere had accumulated. It is probable that this environment was infused with an abundant supply of organic molecules, as the result of both exogenous delivery and endogenous production. The processes and mechanisms that transformed the prebiotic world into a biotic one within the next 500 million years (section 1.3) is at the core of research on the origin of life. As noted at the outset in chapter 1, two distinct methodologies are used to address the question: chemical evolution from simple to complex (bottom up), and the phylogenetic search for the earliest common ancestor to modern life (top down). This chapter reviews the current status of research, considering each of these approaches, and assesses how they might ultimately converge.

Water is the ideal solvent for biochemical processes (section 8.3.3), and the rock–water interface is a favorable locus for chemical evolution (section 5.4.2) that aids catalytic formation and polymerization of organic molecules (see Lambert 2015 for a review). In principle, any aqueous surface or subsurface environment might contribute if an energy source and the essential chemical elements and compounds are available over a suitable range of temperatures: possible scenarios include hot springs and other shallow surface water, fissures and cavities where water seeps deep into the Earth's crust, and hydrothermal systems on the ocean floor. Of these, oceanic hydrothermal systems (figure 6.1) are especially promising, as they give rise to conditions that sustain not only chemical evolution but also living ecosystems, in which chemosynthetic organisms flourish in the absence of sunlight in a setting that has a degree of immunity to conditions on the surface.

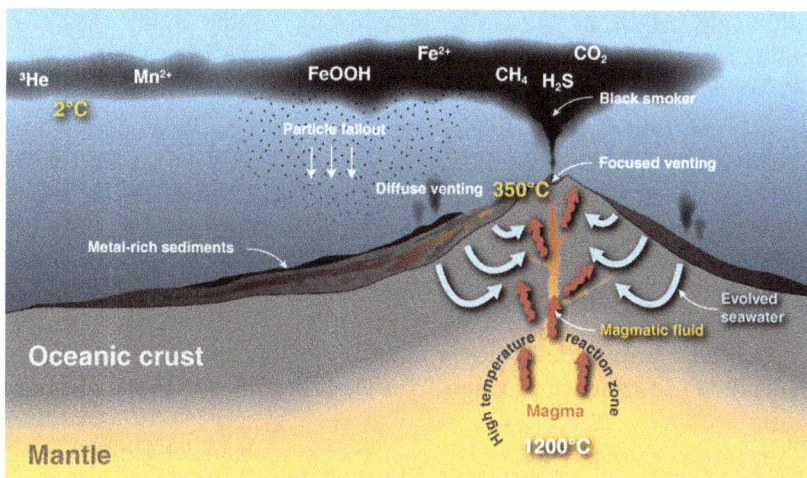

Figure 6.1. Schematic illustration of a hydrothermal system on the ocean floor. Circulation of water through the system is fed by seepage through the crust and driven by heat from rising magma, culminating in high-pressure vents where hot, chemically-enriched fluids flow into cold seawater. These environments promote abiotic chemical evolution and support chemosynthetic microbial communities. Image credit: G J Massoth and C E J de Ronde, GNS Science.

6.1 Constructing life's framework

In this section we consider the bottom-up approach to the origin of life: chemical pathways that might lead from simple molecular precursors to functional biomolecules and biochemical processes on the early Earth (see section 1.4 for a general introduction). Key problems that need to be addressed include the origin of homochiral biomolecules, the emergence of the distinct functions of replication and metabolism, and organization of the basic components into the first living cells.

6.1.1 The RNA world hypothesis

The RNA world hypothesis, introduced in section 1.4.2, proposes a precursor to modern biology in which RNA (ribonucleic acid) acts as both a carrier of genetic information and a catalyst of chemical reactions. The 'chicken and egg' problem of explaining how the distinct but synergistic roles of DNA and enzymes in modern biology could have arisen spontaneously is thus avoided by initially attributing both to RNA. Each individual RNA nucleotide consists of three molecular subunits: a sugar (ribose), a phosphate, and one of four possible nucleobases (G, C, A, U; see figure 1.3). The ordering of the bases in a strand of RNA carries information that can be replicated according to the base-pairing rules, two steps being needed to form an exact copy (e.g. G-A-C-U → C-U-G-A → G-A-C-U). Long RNA strands can fold and cross-link to form diverse structures, and some of these structures have properties that enable them to function as enzymes (ribozymes). In modern biology, ribozymes are a key component of *ribosomes*—complex molecular 'factories' found within all living cells that synthesize proteins by linking amino acids to form polypeptide chains. Peptide synthesis would initially have been far less efficient in the RNA world; but once available, these polymers would have contributed to catalytic activity, leading to production of increasingly complex RNA structures, protein-based enzymes, and ultimately DNA. Proteins and DNA then assume the primary metabolic and genetic roles, respectively, and RNA becomes an intermediary.

Could RNA have formed in sufficient concentrations on the prebiotic Earth for this model to be feasible? Let us consider this question in two stages: (i) production of the subunits (bases, ribose and phosphate) to enable individual RNA monomers to form, and (ii) chirally-selective polymerization of the monomers into RNA strands. Once such strands had been produced, the ability of RNA to autocatalyze might contribute to further production. However, both of the prerequisite steps are problematic.

The three components of an RNA monomer are chemically diverse and form under differing conditions. Nucleobases are relatively straightforward to make by polymerization of HCN in simulated early Earth conditions, as has been demonstrated in the laboratory; indeed, they are present in meteorites, indicating production on asteroidal parent bodies as well. Similarly, sugars may be synthesized by polymerization of H_2CO (the formose reaction), and are also present in trace amounts in meteorites. But for efficient production, the formose reaction requires H_2CO to be present in high (and perhaps unrealistic) concentrations, and in an *alkaline* solution, contrary to the probably somewhat acidic pH state of the early oceans (section 5.4). Depending on the geochemical setting, hydrothermal systems may generate alkaline fluids, and experiments that simulate such conditions demonstrate that ribose is, indeed, produced. However, ribose is typically no more than a minor component of the product mixture (it tends to become 'lost in the sugar forest'; Chyba and McDonald (1995)), and the L and D enantiomers are formed with equal probability. Moreover, free ribose is unstable and subject to decomposition in geochemical environments on timescales of about a year.

The phosphate component is also problematic—indeed, it presents a challenge for any model for the origin of life. P is by far the least abundant of the chemical elements needed to make nucleotides (i.e. in comparison to CHON group; see section 2.1) and its availability thus naturally tends to be a limiting factor. Inorganic phosphates are present in the Earth's crust, typically in the form of apatite, a class of calcium phosphate minerals; however, these minerals are characterized by poor water solubility and low chemical reactivity, so the phosphate groups they contain are not readily accessible for inclusion in biomolecules. Moreover, as *orthophosphates* (PO_4^{3-}), they lack the high-energy P–O–P bonds that characterize polyphosphates utilized in biochemistry. Energized forms such as pyrophosphate ($P_2O_7^{4-}$) may condense from dissolved orthophosphates in hot ($\gtrsim 100$ C), acidic water, conditions consistent with an origin in hydrothermal systems. With a propensity to generate gradients in pH as well as in temperature, pressure and redox state, hydrothermal systems might, perhaps, support conditions under which all three components of RNA nucleotides can form, coexist and combine.

Polymerization of RNA requires both a catalyst and a source of chemical energy, in order to form the strong phosphodiester bonds that link the sugar-phosphate backbone. In living systems the catalysts are biological, and chemical energy is provided by the presence of a diphosphate or triphosphate group in the nucleotide, which is then said to be activated; the additional phosphate units detach and release energy to enable polymerization. How could this occur in the prebiotic world? Minerals such as clays, sulfides, calcite and zeolite have catalytic properties that may simulate the role of biological catalysts in living systems. Their surfaces are not merely passive substrates: they contain active sites in which the specific arrangement of atoms, and the resultant electronic charge distribution, aids the attachment of reactants and facilitates their reaction. Montmorillonite clays, in particular, have been shown in the laboratory to enable polymerization of RNA to produce strands ~50 units long. However, the reaction proceeds only if the monomers are first activated by attaching an additional molecular group (imidazole is typically used in the laboratory). Activation is generally inhibited under prebiotic conditions, and this is a serious hurdle to our understanding of how the first RNA polymers could have formed. The current focus is on laboratory simulations to ascertain whether conditions exist within the range of hydrothermal environments that permit polymerization reactions without prior activation (see, for example, Burcar *et al* 2015).

6.1.2 Chiral selectivity

Homochirality is a defining characteristic of life on Earth (section 1.1) and a successful model must account for its origin. The surfaces of minerals may be implicated: calcite crystals, for example, are inherently chiral, i.e. the arrangement of surface atoms on a given crystal face is not superimposable by its mirror image. The probability of a chiral molecule attaching to the surface may then depend on its handedness, and the products of any chemical reactions that occur on the surface will preserve this imbalance. However, if one surface favors D over L enantiomers, its mirror image (occurring with equal probability) will favor L over D, so the overall chirality of the system is unchanged. Autocatalytic reactions may serve to amplify any localized imbalance, and the enhanced stability and functionality of chirally-pure polymers may give them an advantage over those that are chirally mixed.

Endogenous catalytic processes may thus create localized asymmetries, but they offer no explanation for nature's choice of D over L in nucleic acids and L over D in proteins. The only known source of a potentially planet-wide imbalance is the excess of L-amino acids delivered by meteorites (section 4.3.2). In modern biology, transfer-RNA within the ribosome selects L-amino acids to construct chirally-pure peptide chains. It is interesting to speculate whether, in the RNA world, such selectivity might somehow have been inverted, with an excess of L-amino acids imposing a preference for the D form of RNA.

6.1.3 Metabolism first?

Obstacles to the production of functional RNA polymers under prebiotic conditions, summarized above, naturally prompt an assessment of alternatives to an RNA world, or to consideration of a possible precursor phase. Scenarios that prioritize amino acids over nucleotides as the unit of currency for the first functional biomolecules have an evident advantage in that amino acids are synthesized relatively easily, and may have been delivered in large quantities from space (section 5.4.1) with a pre-existing chiral imbalance of the correct sign (section 4.3.2). As is the case for RNA, polymerization of amino acid monomers necessitates an input of energy and cannot proceed at a significant rate in equilibrium conditions. Experiments indicate that clays and other minerals facilitate abiotic peptide formation, both on the Earth's surface in environments subject to cycling between wet and dry conditions and in hydrothermal systems at higher temperatures. Peptide strands may fold into 3-dimensional structures to form simple proteins, and the primary role of proteins in modern biology is to catalyze metabolic reactions. It is thus suggested that a primitive form of metabolic biochemistry might have preceded the emergence of an RNA world.

The functionality of a protein is determined by its shape, and this is dictated by the sequencing of the amino-acid monomers in the peptide. If sequencing is random and limited to the 20 amino acids used in modern biology, and assuming that a strand at least 10 units long is needed for basic functionality, there is an enormous number ($20^{10} = 10^{13}$) of possible permutations! Life overcomes this problem by using nucleic acid templates to customize the sequencing. If polymerization in the protobiotic world occurred on a mineral substrate, the sequencing may have been influenced by the nature of the surface, such as the repetitive patterns of a crystal lattice, suggesting that such minerals might have served as inorganic templates that favored certain sequences. Organic templates that might have preceded RNA have also been proposed, including peptide nucleic acids (PNA). These polymers resemble RNA in that they contain a sequence of nucleobases attached to a backbone; but in this case the backbone is a peptide chain, formed by polymerization of aminoethyl glycine, which is non-chiral. Experiments have shown that PNA is more stable than RNA and can store and communicate genetic information in an analogous way. Further research is needed to test this hypothesis.

A metabolism-first scenario that has been discussed in the context of hydrothermal environments is the iron–sulfur world hypothesis. Certain protein enzymes in modern biology contain iron–sulfur groups (cofactors) that might be regarded as 'molecular fossils' of an early metabolism in which iron sulfide minerals played a vital role. Hydrothermal systems are rich in sulfides of iron and other transition metals: ferrous iron sulfide (FeS) deposits are primarily responsible for the blackness of the 'black smoker' vents in these systems (figure 6.2). FeS may react with dissolved H_2S within the effluent to form pyrite (FeS_2):

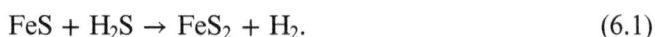

$$FeS + H_2S \rightarrow FeS_2 + H_2. \tag{6.1}$$

This important exothermic reaction provides a source of both free hydrogen and chemical energy, which may then drive a sequence of energetically-favored reactions

Figure 6.2. Examples of two distinct types of hydrothermal vent: a black smoker (left) and a white smoker (right), located at the Brothers and Northwest Eifuku submarine volcanoes, respectively. Black smokers emit acidic, high-temperature (~350 °C) fluids, rich in metal sulfides that condense into black 'smoke' on contact with cold seawater. White smokers are typically situated further from the magma source; their fluids are typically cooler (~250–300 °C), alkaline, and rich in silicon and calcium compounds that yield lighter-hued condensates. Image credits: NOAA.

that reduce CO and CO_2 to form a variety of organic compounds in a sequence of increasing complexity. Suggested products include amino acids, peptides, lipids and the coenzyme acetyl-CoA, a molecule that plays a vital metabolic role in living cells. The dynamic flow of hydrothermal fluids through the system (figure 6.1) ensures a continuous source of reactants, and imposes systematic variations in physical conditions. Acidity gradients resulting from alkaline outflows merging into acidic seawater may facilitate chemical energy storage, e.g. by polyphosphate production, in a manner analogous to a fuel cell. Temperature gradients allow for different reactions to be favored in different regions of the vent, such as monomer synthesis in the lower, hotter regions, followed by polymerization as the fluid rises and cools. The porous mineral walls of the vents may provide favorable locations for products to accumulate and evolve.

6.1.4 Organization and the emergence of cells

The bridge from complex chemistry to primitive biology necessitates a means of organizing the essential ingredients into containers that serve as primitive cells (section 1.4.3). A crucial advantage of the cell (figure 6.3) is the fact that it encloses a microenvironment, distinct from its surroundings, that may be optimized for

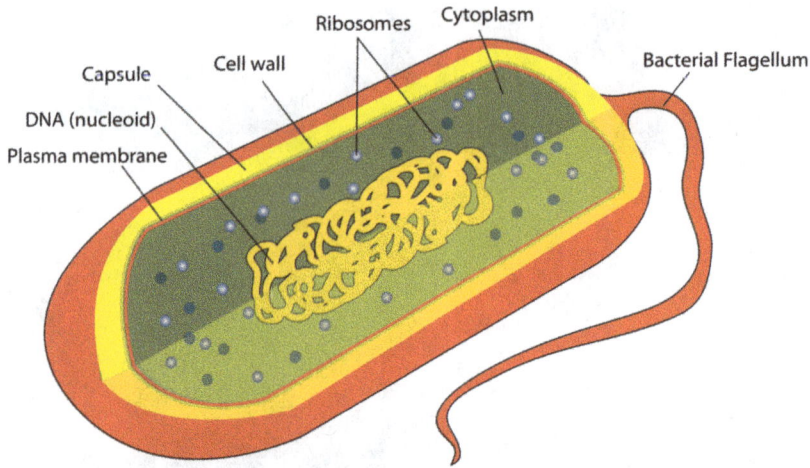

Figure 6.3. Basic structure of a prokaryotic cell.

biological functionality. The membrane in the cell wall is composed primarily of a phospholipid bilayer (figure 6.4), in which the hydrophilic heads and hydrophobic tails of the phospholipids in each layer point outward and inward, respectively, relative to the surface of the membrane. Phospholipids naturally tend to self-organize in water to form bilayers, and a bilayer bubble (a liposome) provides the basic structure of a cell membrane. Additional components such as glycolipids and proteins are included in the membranes of living cells to enhance stability, regulate permeability, and enable communication with other cells.

Phospholipids may have been present on the prebiotic Earth, but perhaps not in sufficient concentrations to be the initiators of cell formation. The voids present within porous minerals are proposed as alternative sites that would have been both plentiful and similarly advantageous. Possibilities include not only the micrometer-sized cavities within porous minerals such as feldspars but also the inter-layer spaces within layered minerals such as mica and clays. Such minerals are common and naturally abundant at the rock–water interface in both surface and subsurface environments where prebiotic chemistry is expected to occur. They provide relatively stable environments for chemical reactions to proceed, perhaps accelerated by catalytic properties of the mineral itself, in enclosures protected from dilution of the products.

In summary, voids within porous minerals appear to be logical and feasible sites for compartmentalization of the first biomolecules and biochemical processes on the early Earth. If this was the case, however, the mechanism by which emergent life detached itself from its geochemical birthplace to form independent cells, bounded by phospholipid membranes and capable of self-replication, remains an open question.

6.2 Tracing life's ancestry

6.2.1 The phylogenetic tree

The top-down approach to origins of life research is based primarily on phylogenetics: studying the evolutionary history of relationships between biological

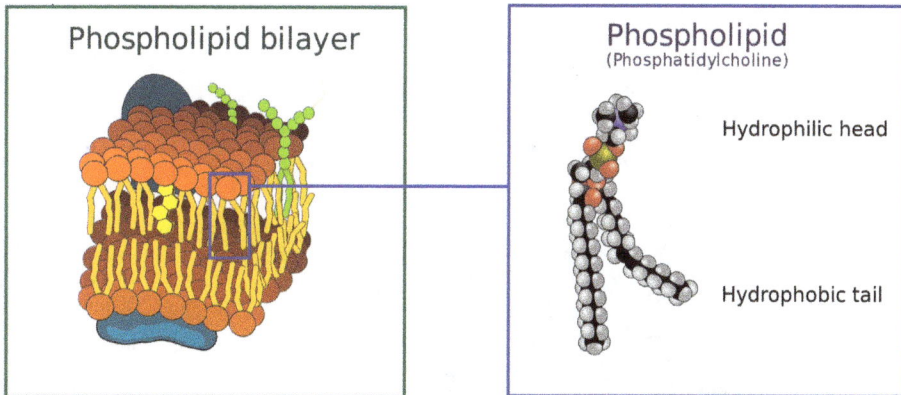

Figure 6.4. Schematic illustrations of a phospholipid bilayer in a cell membrane (left), and an individual phospholipid (right). The membrane also contains glycolipids (green), cholesterol (yellow) and proteins (blue) that contribute to its stability and functionality. Individual atoms in the phospholipid molecule are color-coded grey (H), black (C), blue (N), red (O), and yellow (P). Image credits: D Hatfield and Mariana Ruiz (Wikimedia Commons), edited by the author.

organisms. These relationships are explored by investigating inheritable traits, an idea that dates back to Charles Darwin's *Origin of Species*. Morphological characteristics are observed and compared in both extant life and the fossil record. Modern phylogenetics includes the study of inheritable characteristics at the molecular level, such as DNA and RNA sequences, and ribosomal structure. A detailed discussion of phylogenetic methods is beyond the scope of this book: this section summarizes key results and assesses their implications. See Hedges and Kumar (2009) for in-depth discussion.

The phylogenetic tree (figure 6.5) provides an overview of the relationships between groups of organisms. The groups are classified into three domains, each represented by a major branch of the tree. Bacteria and archaea are the simplest lifeforms (both are unicellular prokaryotes), of which archaea are considered to be somewhat more evolved. Eukaryotes, distinguished by the presence of a well-defined nucleus within each cell, are more complex and may be either unicellular or multicellular. The significance of the phylogenetic tree for the origins of life is the opportunity to follow its branches back toward the root, in an attempt to identify the hypothetical last universal common ancestor (LUCA), i.e. the most recent common ancestor of all current life on Earth. Note that LUCA was not necessarily the *first* life, but the earliest to which an ancestral link may be established. Complications that must be taken into account include the likelihood that the genetic record is incomplete (branches may be missing), and the possibility of horizontal gene transfer (the branches may interconnect). These caveats limit what can be learned regarding the origin of life.

Nevertheless, a significant result is clear: the earliest ancestors that can be identified with confidence are anaerobic hyperthermophiles—organisms adapted to life at extremes of high temperature ($\gtrsim 80$ C). This finding has two possible interpretations. One might simply conclude that life arose in high-temperature

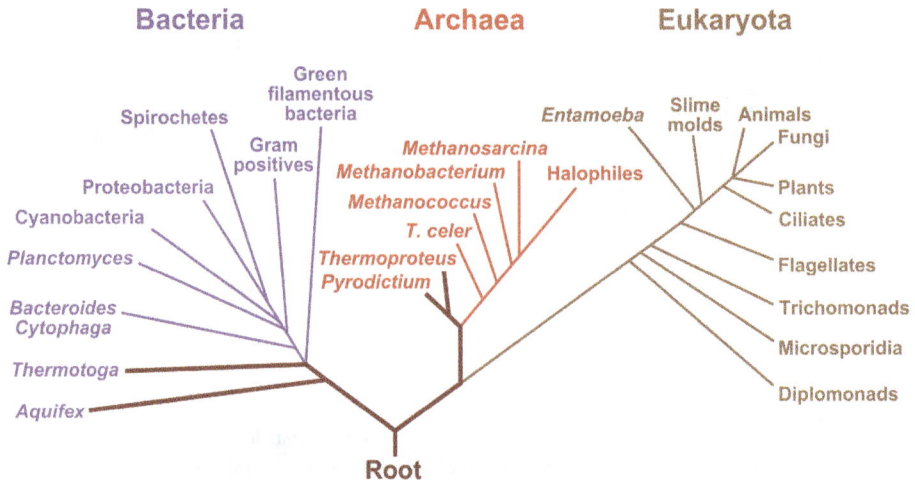

Figure 6.5. Simplified phylogenetic tree of terrestrial life, illustrating the distribution of the three domains (bacteria, archaea and eukaryota). Linkages between hyperthermophilic organisms are highlighted in dark red. The root of the tree suggests an apparent origin from which all life might have evolved. Image credit: NASA Astrobiology Institute.

environments: this would, indeed, be expected if it formed in hydrothermal systems—environments well suited to chemical evolution toward life, as discussed in the previous section, and which continue to support life today. But there is another possibility, arising from the fact that hyperthermophiles, especially those residing deep in the oceans, are fittest to survive the devastation of a large cosmic impact (section 5.2.1). So it is possible that life originated under less extreme conditions, and subsequently diversified. A large impact, perhaps associated with the late heavy bombardment (section 5.2.3), may then have caused global extinctions, leaving hyperthermophiles as the sole survivors. Phylogenetics cannot distinguish between these possibilities.

It is important to determine the chronology of the phylogenetic tree as well as its structure. When did the various branching points between different species occur? And how old are the oldest species? The chronology is constrained by the fossil record, augmented by a technique that uses 'molecular clocks'—the mutation rates of biomolecules—to deduce the timings of divergences between species. The precision to which this can be accomplished declines as we look further back in time. Some of the oldest microfossils, dated at 3.5 Ga, appear to be cyanobacteria, a species still common on the Earth today. As cyanobacteria are several branches above the root of the tree (figure 6.5), significant evolution must have occurred before then, so LUCA must date back to an earlier time—perhaps as early as 4.2 Ga, according to Hedges and Kumar. It seems clear that LUCA was a bacterial species, and that the divergence leading to archaea quickly followed, whereas the first eukaryotes did not appear until later, perhaps about 3 Ga ago. The bacteria and archaea domains both include hyperthermophiles, whereas the eukaryota domain does not (figure 6.5). Additional support for an early origin in a hydrothermal setting arises from studies of sedimentary rocks from the Nuvvuagittuq Greenstone Belt in Quebec, Canada, which contain structures that might be interpreted as fossil remnants of ancient bacteria, resembling those found in modern hydrothermal systems (Dodd *et al* 2017). The chronology is uncertain because the sediments include mineral components with different radiometric ages, but they probably date back to at least 3.77 Ga and possibly as far as 4.28 Ga before present.

6.2.2 Life's early imprint

Once life became established it began to exert a planet-wide influence. If it originated in high-temperature and/or deep-ocean environments, it had clearly diversified by 3.5 Ga to occupy shallow surface water and tidal zones at more moderate temperatures. This is evident from studies of stromatolites dating from this period (figure 6.6). These structures are the fossilized remnants of ancient microbial mats, occupied by colonies of bacteria and archaea, including cyanobacteria, which are photosynthetic. Thus, life seems to have quickly established alternative means of food production from available energy sources as it adapted to different environments. In the absence of sunlight, deep-ocean organisms use chemical energy to convert CO_2 into carbohydrates (chemosynthesis), for example

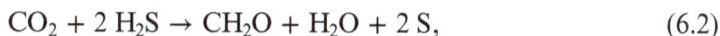

$$CO_2 + 2\,H_2S \rightarrow CH_2O + H_2O + 2\,S, \tag{6.2}$$

a reaction that represents just one of many possibilities. In contrast, the well-known photosynthesis reaction driven by sunlight releases free oxygen into the atmosphere:

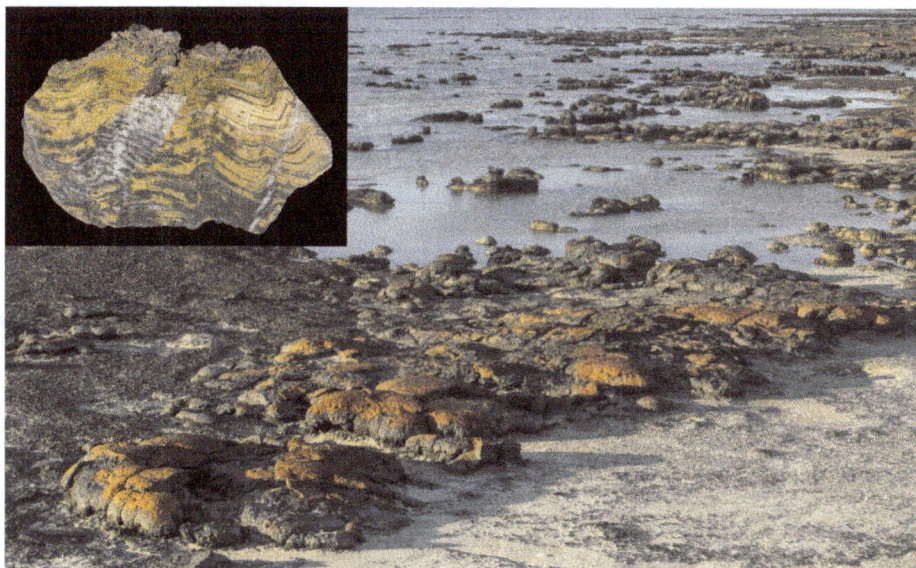

Figure 6.6. Stromatolites, ancient and modern. Main image: present-day stromatolites in shallow coastal water at Shark Bay, Western Australia. Inset: a section through an ancient (~3.5 Ga) fossilized stromatolite from Pilbara, Western Australia (25 × 15 cm in size), showing multilayered sheet structure typical of microbial mats. Image credits: Martin Kraft (main image) and Didier Descouens (inset), Wikimedia Commons.

$$CO_2 + H_2O \rightarrow CH_2O + O_2. \tag{6.3}$$

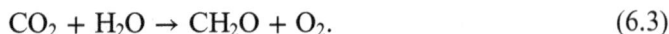

Transformation of the atmosphere from its anaerobic, CO_2-rich beginnings to its aerobic modern state is a topic of great interest, not only in the context of studying the Earth itself but also because it informs our search for a potential biosignature on earthlike planets elsewhere.

The geological record provides insight into the rise of atmospheric oxygen. Banded iron formations—iron–oxide layers within sedimentary rocks—are thought to have resulted from oxidation of Fe dissolved in seawater. Being insoluble in water, the oxides solidified and accumulated on the seabed, forming sediments that were eventually elevated above sea level. The oldest known examples are dated at $\sim$3.75 Ga, and it is therefore logical to presume that the free oxygen needed for their production was released by early photosynthetic organisms such as cyanobacteria. However, the large majority of banded iron formations were laid down more recently (1.9–2.4 Ga ago), implying a long delay, perhaps as long as 1.5 Ga, between the initiation of photosynthesis and the so-called great oxidation event (see figure 1.1), at which point atmospheric O_2 levels began to approach modern values. This delay is readily explained as the timescale for saturation of all such oxygen sinks on the early Earth, including Fe and other metals, on the solid surface and in the oceans. Only when this had occurred could atmospheric O_2 accumulate in large quantities. By then, as William Schopf has remarked, 'lowly cyanobacteria—pond scum—had rusted the world!'

6.3 The transition to life

A true understanding of the origin of life requires convergence of the bottom-up and top-down methodologies, in order to bridge the gap between prebiotic chemistry and extant biology. As Peters and Williams (2012) have emphasized, there should be no discontinuity between the prebiotic and the biotic: each step that led from one to the other must have been thermodynamically favored under the prevailing conditions. Progress has been made toward understanding this transition and the conditions under which it occurred. Environments favorable to prebiotic chemical evolution have been identified at the rock–water interface in hydrothermal settings. It may be no coincidence that the earliest microbial ancestors to modern life that we can currently identify were adapted to function in these same conditions. Elements of the prebiotic world, such as metallic cofactors in protein enzymes, have been identified that were incorporated into living systems and retained in extant biology. Life may thus be linked, by traceable threads, to its prebiotic chemical origins. But many areas of uncertainty remain: some key questions are summarized here.

How long did it take? Can both the timescale and the timing be better constrained? The chronology of life's origin is an important focal point for both bottom-up and top-down methodologies. How rapid was the transition from prebiotic to biotic? And how much time was then available for the first life to evolve into cyanobacteria and other species that colonized stromatolites 3.5 Ga ago? Our understanding of conditions during the earliest periods of Earth's history has increased markedly in recent years (see chapter 5), leading us to conclude that the planet became hospitable to life more quickly than had previously been assumed. Some investigators have suggested, based on phylogenetic models and a tentative interpretation of the rock record (section 6.2), that life may have originated as early as 4.2–4.3 Ga ago. To test this possibility, it will be important to refine techniques for identifying organic matter of biological origin in the oldest samples to avoid possible ambiguities: for example, evidence based on the commonly-used $^{12}C/^{13}C$ isotopic ratio (section 1.3) has been challenged because it might also be effected by abiotic processes.

If life dating back to 4.2–4.3 Ga is confirmed, this raises another question: *Is evidence for the earliest life consistent with the timing of the last planet-sterilizing impact?* If the usual interpretation of the lunar cratering record is correct, such impacts may have occurred as recently as 3.8–3.9 Ga (section 5.2.3). The apparent discrepancy between these timings may be only marginally significant, given the uncertainties, but nevertheless it is important to refine them. Doubt has been cast on the timing, the severity, and even the reality of the so-called late heavy bombardment, and this needs to be resolved. Another challenging problem is to distinguish between the last impact that eradicated *all* life on the planet and the last that eradicated *all except* hyperthermophiles residing in the deepest regions of the biosphere. If robust new evidence for very early life on Earth is found, this may inform our interpretation of the impact threat, rather than the other way around.

How did RNA form under prebiotic conditions? The prominence and longevity of the RNA world hypothesis in origins of life research seems remarkable, given that no realistic pathway has been identified for abiotic formation and polymerization of this molecule! The problem is discussed in some detail in section 6.1.1 above. Hydrothermal systems may be the most promising locations for production to occur, a possibility that might be verifiable experimentally. A better understanding of how and where RNA formed might then clarify the question of whether the RNA world was preceded by another protobiotic chemical system, perhaps dominated by abiotic peptide synthesis and simple metabolic cycles, or whether these processes were a consequence of the RNA world. Finally, the question of how DNA emerged to become the repository of genetic information needs to be clarified. This step seems logical, given the greater stability of DNA compared with RNA, but the process is not yet understood.

Can the origin of life from abiotic precursors be demonstrated in the laboratory? To simulate the entire process would, of course, be the ultimate goal. An evident and seemingly insuperable hurdle is the fact that nature's timescale for the process is unknown and presumably lengthy: even if (as seems increasingly likely) it was rather rapid by astronomical and geological standards, it may still have taken a 100 million years! The only realistic strategy is to demonstrate the plausibility of individual steps in a logical sequence. Indeed, such steps are the focus of much current research. The main challenge is to identify the most realistic set of conditions, and to simulate them in the laboratory. Simulations of deep-ocean hydrothermal environments, in particular, are challenging but also potentially rewarding. Experiments in both prebiotic chemistry and biotic functionality under realistic early-Earth conditions may narrow the divide between the two.

Finally, given that nature evidently did find a way to make life from abiotic precursors, we may question: *Does the entire process still operate on Earth today?* The answer is probably not. Extant life has developed fast, efficient mechanisms to sustain and replicate itself that render the abiotic pathways that led to its origin obsolete. Moreover, life has become pervasive: it occupies all habitable environments on the planet, from deep oceans to the rocky surface and atmosphere, from hot springs to arctic ice; and it has profoundly altered the environment planet-wide, most notably by rendering it generally aerobic. Even if some of the steps that generated relevant products still operate in certain niche environments, those products are likely to feed into extant life rather than the next abiotic step toward a new origin.

Questions and discussion topics

- What are the main points in favor of and against the RNA world hypothesis?
- Consider what research might be done to determine whether a 'metabolic world' preceded the RNA world.
- Which is the least abundant of all the chemical elements needed to make RNA? Is there a problem utilizing this element in prebiotic synthesis?
- Consider whether any alternatives to photosynthesis as sources of free oxygen might have existed on the early Earth.
- Consider what relevance (if any) the lack of an ozone layer in the atmosphere of the early Earth might have had to the origin and early evolution of life.
- What research might be done to clarify whether hyperthermophiles near the root of the tree of life were the first life forms to exist or survivors of a global extinction event?

References and further reading

Barge L M *et al* 2014 The fuel cell model of abiogenesis: a new approach to origin-of-life simulations *Astrobiology* **14** 254

Bell E A, Boehnke P, Harrison T M and Mao W L 2015 Potentially biogenic carbon preserved in a 4.1 billion-year-old zircon *Proc. Natl. Acad. Sci.* **112** 14518

Burcar B T, Jawed M, Shah H and McGown L B 2015 *In situ* imidazole activation of ribonucleotides for abiotic RNA oligomerization reactions *Orig. Life Evol. Biosph.* **45** 31

Chyba C F and McDonald G D 1995 The origin of life in the Solar System: Current issues *Ann. Rev. Earth Planet. Sci.* **23** 215

Dodd M S *et al* 2017 Evidence for early life in Earthas oldest hydrothermal vent precipitates *Nature* **543** 60

Hedges S B and Kumar S (ed) 2009 *The Timetree of Life* (Oxford University Press)

Lambert J-F 2015 Origins of life: From the mineral to the biochemical world *BIO Web of Conf.* **4** 12

Martin W, Baross J, Kelley D and Russell M J 2008 Hydrothermal vents and the origin of life *Nat. Rev. Microbiol.* **6** 805

Peters J W and Williams L D 2012 The origin of life: Look up and look down *Astrobiology* **12** 1087

Schopf J W 1992 The oldest fossils and what they mean *Major Events in the History of Life* (Jones and Bartlett) p 29

Chapter 7

The search for life on Mars

The prospect of life on Mars is a topic with a long history. It has captured the public imagination for more than a century, inspiring the efforts of science fiction writers and movie makers as well as scientists, educators, space program engineers and administrators. Our view of the red planet has changed dramatically over the years. The idea advocated by Percival Lowell and others of an advanced Martian civilization that built a planet-wide system of irrigation canals[1] was quickly dispelled by the first results from the space program. The advent of space-based observatories and explorer missions has radically changed our view of Mars (e.g. figure 7.1), our observations no longer limited to ground-based telescopes peering through the Earth's turbulent atmosphere. Temperatures and pressures turned out to be too low to support liquid water on the surface of Mars today. Nevertheless, space images revealed that channels are, indeed, present on Mars—now dry, but apparently etched in the past by the flow of surface water. This and other evidence suggests that Mars was much more hospitable to life at earlier times.

If life did form on Mars it may have become extinct, or it may still survive at subsistence level in some niche environment, most probably underground. The current search considers both of these possibilities, with a focus on biosignatures of microbial life, past or present. Extremophiles—organisms capable of surviving in the harshest of environments—may have adapted to gradually worsening conditions. Potential terrestrial analogs include cryophiles (adapted to low temperature), xerophiles (adapted to minimal water), and endoliths (adapted to life within porous crustal rocks). A strain that combines all of these traits might be the ideal modern Martian. This chapter reviews what is known of evolving conditions on Mars that

---

[1] In the late 19th century, Italian astronomer Giovanni Schiaparelli had reported observations of 'canali' (meaning 'channels' but easily mistranslated as 'canals'). They were evidently illusory, as they do not correspond to the Martian dry channels imaged by modern spacecraft, which cannot be seen telescopically from Earth. In other respects Schiaparelli's drawings were accurate, and many of the names he assigned to prominent features have been retained.

Figure 7.1. Images of Mars taken with the Hubble Space Telescope's Wide Field and Planetary Camera 2. A complete diurnal cycle is shown (the planet has rotated 90° left to right in each successive image compared with the previous one). Labels refer to major areas near the center of each image (see figure 7.2). The northern polar ice cap, prominent in each image, was relatively small at the time, as the observations were made during the Martian northern-hemisphere summer (the southern polar ice cap was tilted away from the line of sight). Other white features are atmospheric clouds composed of ice particles. Although thin, the Martian atmosphere is dynamic and subject to storms, such as the cyclone visible to the left of the ice cap in the Acidalia image. Image credit: NASA/ESA Hubble Space Telescope.

might have have enabled life to gain a foothold, and suggests methods that may enable us to detect it.

7.1 Evolving conditions and habitability

7.1.1 Mars and Earth compared

Table 7.1 lists bulk properties and orbital parameters of Mars in comparison to those of the Earth. Both planets formed inside the frost line of the Sun's protoplanetary disk (section 4.2), and are therefore expected to have been broadly similar in initial composition. The lower bulk density of Mars may reflect relative differences in the availability of rocky versus metallic solids in their accretion zones, as well as differences in gravitational compression (the uncompressed densities of silicate rock and metallic iron are approximately 2700 and 7800 kg m^{-3}, respectively). Factors the two planets have in common include low orbital inclinations (their orbits are nearly coplanar) and very similar spin periods (the length of the day is almost the same). There is also a close correspondence in obliquity, resulting in seasons on Mars analogous to those of Earth, confirmed by the observed waxing and

Table 7.1. Comparison of bulk and orbital properties of Mars and Earth.

Quantity	Mars	Earth	Ratio
Mass (10^{24} kg)	0.642	5.972	0.107
Mean radius (km)	3396	6378	0.532
Mean density (kg m^{-3})	3933	5514	0.713
Escape velocity (km s^{-1})	5.0	11.2	0.45
Mean atmospheric pressure (mb)	7	1000	0.007
Solar irradiance (W m^{-2})	586	1361	0.43
Mean surface temperature (K)	220	287	0.77
Mean solar distance (10^6 km)	227.9	149.6	1.524
Orbital period (days)	687.0	365.25	1.88
Orbit inclination (deg)	1.885	0	—
Orbit eccentricity	0.094	0.017	5.5
Length of day (h)	24.66	24.00	1.027
Obliquity (deg)	25.2	23.4	1.075

waning of the polar ice caps (figure 7.1) throughout the Martian year. The current similarity in obliquity appears to be coincidental, however, as model calculations predict the occurrence of long-term cyclic variations: these variations are minor in the case of the Earth (section 5.1), much larger in the case of Mars, with a predicted range of 15–45 degrees. Another difference affecting global temperatures is the somewhat greater eccentricity of Mars' orbit, resulting in modest annual variations in solar irradiance.

Two major differences that affect planetary habitability (section 4.4) are mean solar distance and mass. Because of its greater solar distance, the mean irradiance (energy per unit area) that Mars receives from the Sun is about 43% relative to that received by the Earth. Perhaps of even greater significance is the fact that the mass of Mars is a mere 11% relative to Earth. The mass determines the escape velocity, which controls a planet's ability to retain an atmosphere (see figure 4.4); it also affects the timescale for persistence of geologic activity (see figure 4.7), which controls its ability to replenish and recycle the atmosphere (section 4.4.2). Both of these differences are consistent with the relatively hostile conditions prevailing on Mars today. The challenge is to understand the evolution of past conditions— conditions that were evidently once favorable to the presence of liquid surface water— in order to assess whether they could also have supported life.

7.1.2 Interpreting the topography

Figure 7.2 shows global topographic maps of the entire Martian surface, color-coded according to elevation[2]. It is immediately evident on inspection that the

---

[2] See Google Mars (http://www.google.com/Mars/) for an interactive map based on the same data.

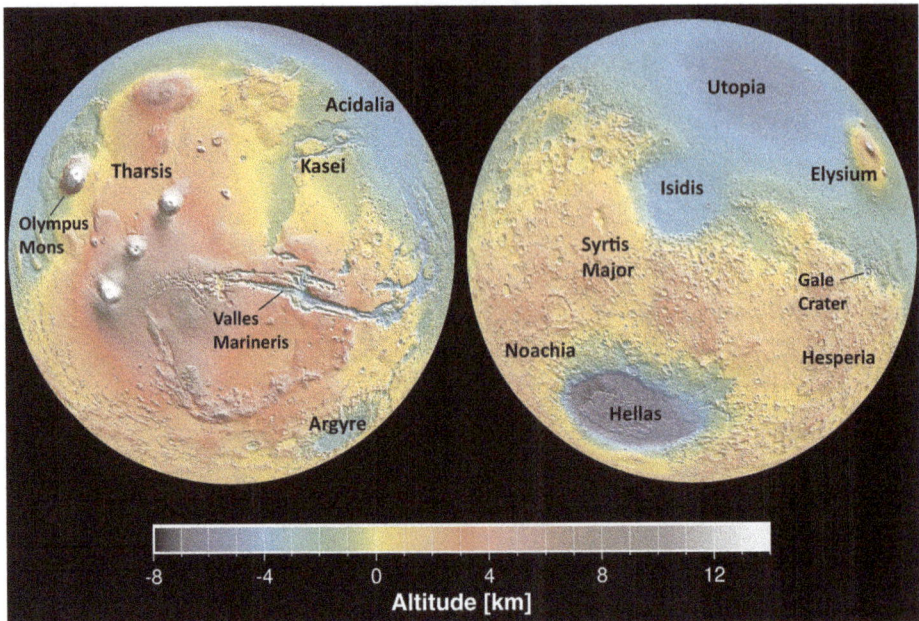

Figure 7.2. Color-coded topographic map of Mars (western and eastern hemispheres) obtained by the Mars Orbiter Laser Altimeter (MOLA). Features mentioned in the text are labeled. The scale indicates elevation relative to an adopted zero level (an equipotential surface normalized to the mean radius of the planet). Image credit: NASA/JPL/MOLA, with labeling by the author.

surface is extremely varied in both elevation and structure, and includes features reminiscent of both the Earth and the Moon:

- Heavily cratered highlands (mostly in the southern hemisphere), such as Noachia and Hesperia, and major impact basins (Hellas and Argyre), resembling impact features seen on the Moon.
- Smooth, lightly cratered plains (mostly in the northern hemisphere), such as Acidalia, Utopia and Isidis, reminiscent of lunar maria.
- Giant volcanic mountains in the Tharsis and Elysium regions, including Olympus Mons, the tallest volcano in the solar system. All of these structures resemble terrestrial shield volcanoes (figure 5.4); none show evidence of current or recent activity.
- Various channel-like features. The largest, Valles Marineris, appears to be a giant rift valley, 4000 km long, 200 km wide and up to 7 km deep. Others appear to have been carved by water. An enlargement illustrating such channels in the Kasei region is shown in figure 7.3.

These features provide insight into the geological history of Mars. By analogy with their lunar counterparts (section 5.2.2), the cratered highlands are the oldest surviving regions of the crust, little altered since the last epoch of heavy bombardment, whereas the lightly cratered plains are perhaps around 1 Ga younger. Other features, such as the rift valley and the clusters of tall volcanoes, are more suggestive

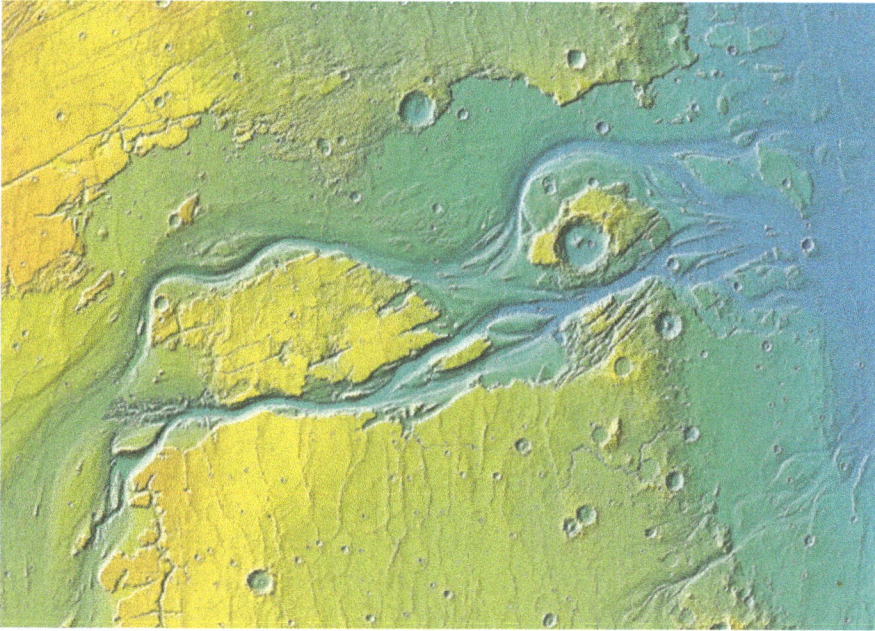

Figure 7.3. Enlargement of the color-coded topographic map (figure 7.2) for a section of the Kasei region, approximately 1600 km wide, showing valleys carved by water flowing eastward from highlands (lower left) to the Acidalia plains (right). Image credit: NASA/JPL/MOLA.

of earthlike geologic activity. A global system of tectonic plate motion is ruled out, however, by the survival of large areas of ancient crust, and by the absence of volcanoes of the *strato* type (see section 5.3.1), indicating a general lack of crustal subduction. The sheer size of the Martian shield volcanoes is also consistent with a static crust. On Earth, a hot-spot feeds magma into a sequence of volcanoes as the plate slides over it (the Hawaiian Island chain being a prime example), whereas on Mars it would have fed continuously into a single structure.

Correct interpretation of the channel-like features on Mars is vital to our assessment of the planet as a potential host to life. Some may be volcanic rather than aqueous in origin, others may be aeolian (shaped by wind rather than water). Nevertheless, many features bear a clear and striking resemblance to terrestrial counterparts formed by the interaction of rock and water: these include not only river-like channels (e.g. figure 7.3) but also 'crater-lakes' such Gale Crater (the landing site of NASA's Curiosity mission), which contains sedimentary layers consistent with an aqueous origin. Geochemical analyses carried out by both orbiting and landed spacecraft provide corroboratory evidence: minerals such as hematite (Fe_2O_3) and clays, which form in the presence of water on Earth, are common on the surface of Mars.

River systems such as the Kasei delta would have drained water into the lower-lying plains. This has led to speculation that a vast northern ocean may once have covered the Acidalia, Utopia and Isidis regions—the choice of blue to represent the elevation of these regions in the topographic map (figure 7.2) may not be coincidental! The smoothness of these plains is consistent with an interpretation as dried-up ocean beds. However, climatological models have difficulty explaining how conditions could have arisen in which such large bodies of surface water remained stable on early Mars.

7.1.3 Evolution of the Martian climate

What were conditions like on early Mars, around the time that life was beginning to emerge on Earth? Could Mars also have been warm and wet, with conditions potentially conducive to an origin of life? Assuming that H_2O was plentiful on early Mars, the presence of the liquid form would have depended on the atmospheric temperature and pressure. On present-day Mars, these are almost always too low, as illustrated in figure 7.4. At early times, the situation would have been exacerbated by the fact that the Sun was significantly less luminous (section 4.4.1). Heat resulting from asteroidal impacts would have caused temporary, localized melting, but global temperatures high enough to support widespread liquid water would have required a denser atmosphere and a substantial greenhouse effect. The Martian volcanoes provide a natural source of such an atmosphere: like their terrestrial counterparts, they must have outgassed large volumes of greenhouse gases such as water vapor and CO_2 during periods of intense activity. Precipitation of H_2O may have ensued in much the same way as on Earth (section 5.3). Atmospheric pressures reached during

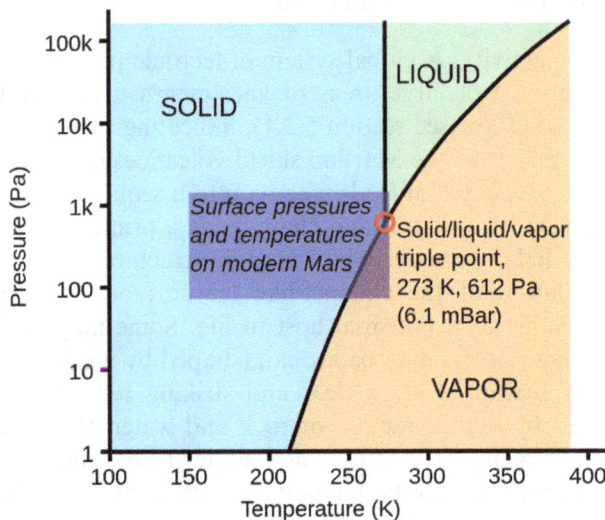

Figure 7.4. Pressure–temperature phase diagram for H_2O, illustrating the range of values prevailing on the surface of present-day Mars (blue rectangle). The range in atmospheric pressure corresponds to the range in elevations, from the deepest basin (Hellas) to the tallest mountain (Olympus Mons); the range in temperature corresponds to different latitudes and seasons. In general, only phase changes between solid and gaseous states occur on Mars today (e.g. seasonal dissipation of polar ice results from sublimation rather than melting). Both mean pressures and temperatures are presumed to have been significantly higher ~3.5 Ga ago, at which time conditions on Mars would have overlapped the liquid water zone. Image credit: Daniel Hobley, Wikimedia Commons.

this period are unknown—we may speculate that they approached or even exceeded terrestrial levels. Because of this, it is not yet clear whether Mars acquired an extensive, global hydrosphere replete with oceans; but the evidence for a 'warm and wet' phase seems compelling, strongly supported by the topographical and geo-chemical evidence discussed above, and qualitatively consistent with what is surmised about the volcanic history of the planet.

The timing of Mars' warm and wet phase may be assessed by comparing the distributions of water features and impact craters, especially in areas where they overlap. In figure 7.3, for example, it may be discerned that the ejecta surrounding the prominent impact crater (Sharonov) to the right of center has been breached and partly eroded by the channel to its south, implying that the crater formed before the channel, whereas craters overlapping the channel bed are all much smaller. From this and similar evidence from other regions, it is deduced that the water features postdate the last period of intense bombardment, and that only light residual cratering has occurred since the water receded (see section 5.2.2 and figure 5.2). Our ability to convert this information into a secure time-frame is limited, however, by the fact that we do not yet have samples from these regions of Mars that can be analyzed to yield absolute ages. If the impact chronology of Mars followed a similar pattern to that of the Moon, the warm and wet phase occurred approximately 3.2–3.8 Ga ago (the range may extend to earlier times if Mars did not suffer a late heavy bombardment). If Mars did host an origin of life, this is the most likely time period for it to have occurred.

Subsequent evolution gradually transformed the Martian climate from warm and wet to cold and dry. This outcome resulted from the combination of atmospheric loss and declining volcanic activity, both of which are consequences of Mars' relatively small mass, as previously discussed. The natural leakage of atmospheric gases into space that results from their thermal motion (figure 4.4) would have been exacerbated by the Sun's radiation and solar wind. Light gases such as H_2O, NH_3 and CH_4 are all subject to loss, and even N_2 is only marginally stable (nitrogen in the present Martian atmosphere has an unusually high $^{15}N/^{14}N$ ratio, indicating better retention of the heavier isotope). Lacking an ozone layer, the gradually thinning atmosphere became effectively transparent to solar UV radiation, such that even molecular gases near the surface were subject to ionization and dissociation. In the case of H_2O, dissociation resulted in rapid loss of free hydrogen and an accumu-lation of free oxygen. O_2 is no more than a trace constituent ($\sim0.1\%$) of the present-day Martian atmosphere, but the surface is rich in iron oxides (responsible for the planet's distinctive color). Unlike its earthly sibling (section 6.2.2), rusting of the Martian world did not necessarily depend on a biological source of O_2.

CO_2 is presumed to have been a major constituent of an early Martian atmosphere dense enough to produce extensive global warming, a mere vestige of which still remains (the tenuous modern atmosphere is 96% CO_2). To explain the low present-day pressure ($<1\%$ of that on Earth) we must account for a major decline in CO_2 as well as in other atmospheric gases, and this is problematic. The polar regions contain frozen CO_2 in seasonal exchange with the atmosphere, but if all known deposits of solid CO_2 were returned to the atmosphere the mean pressure would increase by only a factor of about two. A much greater potential sink for atmospheric CO_2 is absorption into the crust to form carbonate minerals, a process that occurs extensively on Earth. Carbonates have, indeed, been identified on Mars, but not so far in concentrations sufficient to account for the expected degree of loss from the early atmosphere. Perhaps leakage into space depleted CO_2 as well as the lighter gases: this might have resulted from photodissociation and subsequent loss of the constituent atoms, accelerated by collisions with solar wind particles.

Mars did not lose all of its H_2O during this transformation. Some is present in the polar caps, but the most extensive reservoirs appear to be buried underground. Searches for these reservoirs currently depend on radar techniques, using a sounding instrument carried by the Mars Reconnaissance Orbiter. Transmitted radio waves are partially reflected from surface and subsurface layers, up to a depth of a few hundred meters; analysis of the reflections enables the optical properties of the materials in the layers to be measured, and this yields information on their composition. Structures consistent with subsurface ice have been mapped. They include buried glacier-like features in both hemispheres, and a large body of subsurface ice in the Utopia region, estimated to contain as much H_2O as the Earth's Lake Superior (Stuurman *et al* 2016).

7.2 Searching for biosignatures

The search for evidence of Martian life can be said to have begun in earnest with the Viking program, which delivered robotic experiments designed to detect organic molecules and extant life at two locations on the Martian surface in 1976. Since then, the search has extended to Earth-based laboratory studies of Martian meteorites, and to detailed spectroscopic surveys of the Martian surface and atmosphere by orbiting spacecraft, as well as further *in situ* experiments[3]. It is informative to consider these developments in chronological order, to illustrate the evolution of ideas on how best to conduct the search.

7.2.1 The Viking biology experiments

In 1976 the Viking landers conducted *in situ* experiments at two well-separated locations, one on the Acidalia plains near the Kasei delta, the other on the Utopia plains northwest of Elysium—both low-lying regions that may once have been covered by water. In each case, the samples consisted of loose surface material (regolith) scooped up from areas close to the landing site. The experiments may be summarized as follows.

- A gas chromatography-mass spectrometry experiment, designed to detect molecules present in the regolith. A sensitive search was made for organic molecules.
- A suite of experiments designed to detect the products of biological processes that might be attributed to microbial life in the regolith. To test for

---

[3] Mars has been under continuous investigation by active orbiting or landed space missions since the Pathfinder lander touched down in 1997.

metabolism, samples were mixed with water and organic nutrients, the latter 'labeled' with the radioactive ^{14}C isotope; gases released by the mixture were then analysed for the presence of $^{14}CO_2$ as evidence that the nutrients had been metabolized. To test for photosynthesis, samples were mixed with water in an atmosphere containing $^{14}CO_2$ and exposed to visible light; the sample was then analysed for organic products that had incorporated ^{14}C from the atmosphere.

The results were intriguingly inconsistent. The labeled release experiment gave an apparently positive result for metabolism, the other biological experiments gave negative results, and the gas chromatography-mass spectrometry experiment failed to detect organic molecules, setting upper limits of a few parts per billion in the samples. Much controversy ensued, but a general consensus was reached that life had not been detected, and that the labeled release experiment had given a false-positive result.

This situation arose because the nature of the Martian regolith was not sufficiently well understood when the labeled release experiment was designed. The instrument had operated correctly and the experiment was capable of yielding a true positive result on Earth. However, the loose surface material on Mars (mostly windblown dust) is now known to be very highly *oxidized*. Solar UV radiation, strongly attenuated in the Earth's upper atmosphere, penetrates to the lower atmosphere and surface of Mars, and this generates a supply of free atomic and ionic O that oxidizes surface materials, as noted in section 7.1.3 above. When these materials were mixed with organic compounds, they oxidized the carbon to CO_2, mimicking metabolism. Any naturally-occurring organic molecules are subject to the same outcome, which may explain the dearth of organics in the Martian regolith.

Nevertheless, the Viking mission was a success: it provided a wealth of valuable data and important guidance for the design of future missions. A lesson learned was that the dusty regolith that blankets much of the Martian surface is unhelpful in the quest for biosignatures. Indeed, experiments conducted by the Phoenix lander in 2007 showed that it contains perchlorate (ClO_4^-) compounds, which are toxic to terrestrial life. The Martian atmosphere undergoes periodic global dust storms that transport and distribute this material over large distances. Thus, the regolith is not only devoid of organics and toxic to life but also mostly unrepresentative of native materials at a chosen landing site. The need for a change of focus to crustal rocks and subsurface material was indicated. Serendipitously, data from the Viking mission also provided the method for identification of the first rocks from Mars to be studied in laboratories on Earth: the Martian meteorites.

7.2.2 Martian meteorites

It was realized in the early 1980s that the meteorite collection includes pieces of Mars. Members of this group are distinctive in both composition and age: whereas the large majority of meteorites originated in the solar nebula 4.56 Ga ago (section 4.3.2), those now recognised as Martian formed more recently (most have ages

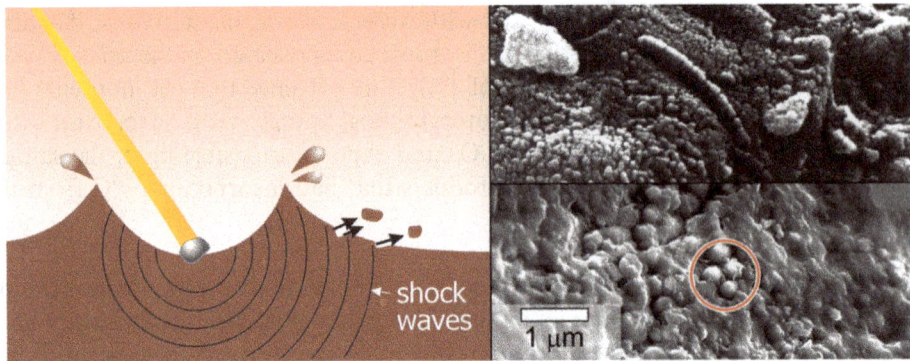

Figure 7.5. Martian meteorites. Left: Schematic illustration of the launch of crustal material during a crater-forming impact on Mars. The energy released upon collision typically melts or vaporizes material close to the point of impact, but material in peripheral regions may be accelerated to escape velocity by the ensuing shock waves with little damage. Martian meteorites are pieces of this material that fall to Earth. Right: Scanning electron microscope images of structures within Martian meteorites Alan Hills 84001 (above) and Yamato 000593 (below). See text for discussion. Image credits: Bruce Watson (schematic) and NASA/JPL (meteorite images).

$\lesssim 1$ Ga) and in a planetary setting (most are igneous rocks that have been subjected to weathering). Their provenance was demonstrated by analyses of pockets of gas trapped within glassy inclusions, which closely match the composition of the Martian atmosphere measured by the Viking landers. Their interplanetary migration is thought to have been instigated by shock waves from crater-forming impacts that launched them from the Martian surface, as illustrated in figure 7.5. Martian meteorites account for $\sim 0.2\%$ of all known meteorite falls on Earth.

These meteorites carry information that contributes to our understanding of the geological and climatic history of Mars. In each case, radiometric dating enables the condensation age of the crust from which it originated to be measured, and the level of cosmic ray exposure yields an estimate of the interplanetary transit time of the meteorite, thus providing some chronological context. The condensation ages of the youngest ones suggest that Mars may have been volcanically active 180 Ma ago (more recently than previously thought). Some experienced aqueous conditions on Mars prior to launch, as evidenced by the presence of embedded minerals that are formed by the action of water on igneous rock: examples have been found in which this may have occurred as recently as 620 Ma ago. These investigations are valuable in elucidating the chronology of global conditions on Mars, albeit limited by a lack of geological context, as the point of origin is unknown.

As potential carriers of biosignatures, the Martian meteorites have proven to be controversial and equivocal. A study of the Alan Hills (ALH) 84001 meteorite led to a proposal, published by McKay *et al* in 1996, that it contains microscopic structures consistent with identification as fossilized remnants of microbial Martian life. The rod-shaped structure seen in figure 7.5 (upper right), for example, resembles a bacillus (although, typically 20–100 nm in diameter, such structures are appreciably smaller than terrestrial bacteria). Recovered from Antarctica in 1984, ALH 84001 is amongst the oldest known Martian meteorites (dated at 4.1 Ga), and it evidently experienced aqueous conditions prior to ejection from Mars. It contains magnetite crystals embedded in carbonate structures that are typically associated with biological activity on Earth. It also contains organic molecules in trace amounts, but most of these may be terrestrial contaminants: only polycyclic aromatic hydrocarbons (PAHs) are confidently identified as indigenous to the meteorite. PAHs are expected to be present in interplanetary debris that accumulated on the Martian surface, and are not therefore considered to be indicative of Martian biology.

The primary objection to an assignment of structures such as those in figure 7.5 to fossilized remnants of microbial life is simply that they are not necessarily biological in origin. Similar structures may be produced by abiotic processes (indeed, the same problem has hindered searches for evidence of the earliest terrestrial life). Because of it, the scientific community has not accepted that they represent a secure detection of Martian life. Nevertheless, the controversy led to renewed interest in the red planet and, more generally, in the search for extraterrestrial life.

A lesson learned is that morphology alone cannot be used as an unambiguous means of detecting microbial life. Geochemical evidence may strengthen the case— the magnetite crystals found in ALH 84001are an example, but again it is possible that they were formed abiotically. A more recent study by White *et al* (2014) investigated potential biosignatures in Yamato 000593, a basaltic meteorite originating from a volcanic region of Mars that formed ~1.3 Ga ago. The meteorite includes layers of a mineral (iddingsite) that forms by the action of water on basalt, and contained within these layers are spheroidal features shown in figure 7.5 (lower right). Spheroids such as those within the red circle were shown to contain carbon of potentially biotic origin. As in the case of ALH 84001, the findings are suggestive of biology but not conclusive proof.

7.2.3 Atmospheric methane

Spectroscopic detections of methane (CH_4) in the Martian atmosphere, first reported in 2004, have been made with both Earth-based and Mars-orbiting observatories. Although the concentrations are very low (typically no more than 10 parts per billion), the presence of *any* CH_4 in the Martian atmosphere is surprising because it is subject to dissociation by solar UV radiation on estimated timescales of a few hundred years. Its presence therefore implies an active source. On Earth, the primary source of atmospheric CH_4 is biological, in the form of methanogens—anaerobic organisms that release methane as a byproduct of metabolism. The observations indicate that CH_4 levels on Mars are variable in time and perhaps also in geographical location. The Curiosity rover has been monitoring atmospheric CH_4 levels in Gale Crater since it landed in 2012, and the results are puzzling. For extended periods, Curiosity found no detectable CH_4, but did record a significant transient rise during a period of about two months in 2013/14 (see Webster *et al* 2015).

Interpretation of these observations is challenging. No known mechanism can explain such variations, which require both an intermittent source and a destruction mechanism that operates more rapidly than the predicted photodissociation time-scale. Reaction of CH_4 with oxygen to form CO_2 may be implicated as an additional mechanism for loss, but current atmospheric models are unable to account for the observations. Methanogenic Martian life is just one of a number of possible sources. Others that have been proposed include geochemical processes (such as serpentinization and Fischer–Tropsch reactions; see section 5.4.2), active volcanism, and seasonal release from polar or subsurface ice. Of these, a volcanic source can probably be ruled out, not only because of the lack of evidence for current or recent volcanism on Mars, but also because any such CH_4 emissions would likely be accompanied by larger quantities of other gases, notably SO_2, that are undetected in the Martian atmosphere. Seasonal exchange with ice reservoirs also seems improbable, given that no clear seasonal correlation has been demonstrated. A geochemical source is perhaps the most plausible; a biological source is unconfirmed but not excluded.

7.2.4 Searching the sediments

Perhaps the ideal strategy for an *in situ* search for biosignatures of Martian life would be to drill a deep core sample in an area covered by sediments laid down during an extended period of immersion in water. The core sample would effectively provide a timeline from the deepest (oldest) to the shallowest (youngest) sedimentary layers. This would give insight into climatic history, and a realistic possibility of detecting organic molecules in sediments protected from the hostile present-day surface environment. It would also enable a rigorous search for the presence of both geochemical and morphological evidence of life. However, to drill a core sample on Mars to an adequate depth (perhaps a kilometer or more) and return it for analysis is not a realistic goal with present technology. A feasible alternative that may yield some similar advantages is to explore a location on Mars where sedimentary layers have been exposed naturally. Gale Crater, the landing site of the Mars Science Laboratory (Curiosity) mission, is one such location.

Gale Crater is an impact feature, 150 km in diameter and perhaps ~3.8 Ga old, containing an unusually prominent central peak. In general, the central peak of a large impact crater results from a dynamical rebound as the crater walls slump under gravity in the immediate aftermath of the impact. In the case of Gale Crater, the central mountain (Aeolis Mons, also known as Mount Sharp; figure 7.6) contains a vast deposit of sedimentary layers, apparently superposed on the original central peak. The crater is presumed to have filled with water during the putative 'warm and wet' phase of Martian history (section 7.1.3); sediments laid down over this period may once have filled the crater, but subsequent erosion has partially removed them to expose underlying layers. The gradient in elevation seen in figure 7.6 corresponds to a gradient in mineralogy. The base of the mountain contains clays and hematite, which originate in aqueous conditions, whereas layers at intermediate elevations are richer in salts and sulfates that may represent a drying phase.

At the time of writing the Curiosity mission is still ongoing. Important results to date include measurements of atmospheric methane (section 7.2.3) and the first *in*

Figure 7.6. Aeolis Mons (Mount Sharp), the 5 km-high central peak in Gale Crater, imaged in September 2015 by the Mast Camera of the Mars Curiosity rover. Colors have been adjusted so that rocks appear approximately as they would on Earth. The slopes show evidence of layering, with successive layers laid down under evolving conditions. Image credit: NASA/JPL/Mars Science Laboratory.

situ detection of organic molecules in crustal material. Analyses of samples drilled from a mudstone sediment in Gale Crater revealed the presence of chlorinated hydrocarbons, including chlorobenzene (a benzine ring with an attached Cl atom) and dichloroalkanes (aliphatic hydrocarbons, each with two attached Cl atoms) at the 70–300 parts per billion level (Freissinet *et al* 2015). These molecules are presumed to be the reaction products of Martian chlorine and organic carbon laid down in the sediments. It is not yet known, however, whether the original organics were products of indigenous processes on Mars or simply accumulations from exogenous sources such as meteorites and interplanetary dust particles. Detections of organics in Martian meteorites (section 7.2.2) are similarly ambiguous.

The problem of detecting complex organic molecules that could be precursors to or products of Martian biochemistry thus remains unsolved. This situation may well arise from an inaccessibility of suitable samples rather than a true dearth of such molecules. Exposed sediments are subject to oxidative degradation, as previously discussed. The mudstone samples analyzed by Curiosity were drilled to depths of only a few centimeters.

7.3 Future prospects

Curiosity and its predecessor missions, such as Spirit and Opportunity, were designed primarily to seek evidence for habitable conditions on Mars, rather than to conduct a definitive search for life itself. The capabilities of robotic instruments that can be landed on Mars have evolved dramatically since the time of the Viking mission, but they cannot yet accomplish all of the tasks that are possible in laboratories on Earth. An example is the scanning electron microscope (SEM), capable of resolving fine structural details within a sample (see figure 7.5). Such instruments are typically both large and power-hungry. Two possible solutions to this problem are under active investigation: innovative design of an automated, miniaturized SEM that could be carried by future lander missions, and plans to return samples to Earth for analysis.

A sample return mission may offer the best prospect of a rigorous search for Martian life. Samples should be collected from the most promising sites, as indicated by the results of prior investigations designed to assess habitability, and from subsurface as well as surface locations. They must then be isolated from exposure to the Martian atmosphere, and returned to Earth in pristine condition. The full suite of laboratory techniques could then be applied, including both high-resolution imaging and sensitive compositional analyses designed to detect organic molecules and chemical biosignatures. As was the case with the Apollo samples returned from the Moon, a portion of the payload could be preserved for study at a later date, allowing for the possibility of future advances in instrumentation and techniques.

The design of a sample return mission is challenging. A sophisticated rover, perhaps similar in design to Curiosity, would be needed to identify and collect the samples at the chosen landing site. However, such spacecraft lack the capacity to relaunch and return to Earth—the additional weight and energy requirements would be prohibitive—so at least one additional vehicle would be needed. A feasible strategy would be to have the initial lander collect and store a cache of samples for future collection by a second vehicle, custom-designed for the purpose. One of the goals of NASA's Mars 2020 mission, named for the planned year of launch, is to initiate this process by caching samples for future return.

Eventually, 'life on Mars' may take human form. Another goal of the Mars 2020 mission is to gather data and test technology that will inform the design of future human exploration missions. It presents an opportunity to refine the precision and reliability of the landing technology. Onboard instruments will test methods to produce O_2 from Martian atmospheric CO_2, and collect data to evaluate hazards posed by Martian dust.

Questions and discussion topics

- Why is Mars red?
- Why is there a dearth of complex organic molecules on the surface of Mars? What is the significance of this finding?
- How can the timing of past aqueous conditions on Mars be estimated? What factors limit the accuracy?
- Distinguish between geological processes that form strato and shield volcanoes on Earth. What is the significance of the fact that only shield volcanoes are found on Mars?
- Consider what landing site(s) you would select for a future Mars sample return mission. What is the reasoning for your choice?
- Given that we have meteorites from Mars, is a sample return mission scientifically justified?
- Consider whether future human exploration of Mars is likely to contribute to the search for indigenous Martian life.

References and further reading

Aerts J, Röling W, Elsaesser A and Ehrenfreund P 2014 Biota and biomolecules in extreme environments on Earth: Implications for life detection on Mars *Life* **4** 535

Craddock R A and Greeley R 2009 Minimum estimates of the amount and timing of gases released into the Martian atmosphere from volcanic eruptions *Icarus* **204** 512

Freissinet C *et al* 2015 Organic molecules in the Sheepbed Mudstone, Gale Crater, Mars *J. Geophys. Res. Planets* **120** 495

McKay D S *et al* 1996 Search for past life on Mars: Possible relic biogenic activity in Martian meteorite ALH84001 *Science* **273** 924

Niles P B *et al* 2013 Geochemistry of carbonates on Mars: Implications for climate history and nature of aqueous environments *Space Sci. Rev.* **174** 301

Stuurman C M *et al* 2016 SHARAD detection and characterization of subsurface water ice deposits in Utopia Planitia, Mars *Geophys. Res. Lett.* **43** 9484

Vasavada A 2017 Our changing view of Mars *Phys. Today* **70** 34

Webster C R *et al* 2015 Mars methane detection and variability at Gale crater *Science* **347** 415

White L M, Gibson E K, Thomas-Keprta K, Clemett S J and McKay D S 2014 Putative indigenous carbon-bearing alteration features in Martian meteorite Yamato 000593 *Astrobiology* **14** 170

Origins of Life
A cosmic perspective
Douglas Whittet

Chapter 8

Icy worlds as potential hosts for life

The preceding chapters assess the habitability and biological status of two planets in or near the 'Goldilocks zone' of our planetary system—the Earth and Mars. These are not, however, the only bodies in the system that might have potential for biology. Bodies that formed at greater solar distances, beyond the frost line of the protoplanetary disk (section 4.2.2), were particularly well endowed with a resource vital for life as we know it: H_2O. In solid phase, H_2O was indeed one of the primary ingredients of bodies formed in this region, whereas the Earth appears to have acquired some and perhaps most of its water serendipitously, as a late veneer of cometary and asteroidal material (section 5.3.2).

Water remains frozen on bodies situated beyond the outer rim of the Goldilocks zone unless an alternative to solar radiation is available as an energy source. Aside from the giant planets, the largest members of this group are significantly less massive than Mars, and are therefore presumed to have dissipated much of their intrinsic internal heat by now, for reasons discussed in section 4.4.2. However, in several cases, the gravitational interaction of an icy world with its neighbors provides a continuing source of internal heat, resulting in volcanic activity, surface modification, and the possibility to maintain a subsurface liquid ocean. The Jovian moons Europa and Ganymede and the Saturnian moons Enceladus and Titan, in particular, have been identified as important targets for investigation. That icy worlds such as these may contain potentially habitable environments does not necessarily imply, of course, that life originated there or that they host life today. This chapter presents an overview of the current status and future plans.

8.1 The moons of Jupiter

Jupiter is unique amongst the planets of our solar system in having as many as four large moons (figure 8.1 and table 8.1): only Titan and the Earth's Moon are of comparable size. Easily seen with a small telescope, they were discovered by Galileo in 1610. As a family, these moons resemble a miniature solar system, revolving

Figure 8.1. The four Galilean moons of Jupiter compared, with relative sizes to scale, and displayed in order of mean distance from Jupiter (from left to right): Io, Europa, Ganymede and Callisto. Each body has unique surface features arising from different degrees of internal heating by tidal flexing. Image credit: NASA/JPL.

Table 8.1. Comparison of data for the four Galilean moons of Jupiter.

Quantity	Io	Europa	Ganymede	Callisto
Mass (10^{22} kg)	8.9	4.8	14.8	10.8
Mean radius (km)	1822	1561	2631	2410
Mean density (kg m^{-3})	3528	3013	1942	1834
Mean orbital radius (10^3 km)	422	671	1070	1883
Orbital period (days)	1.8	3.6	7.2	16.7
Orbital period relative to Io	1	2	4	9.3
Orbital eccentricity	0.004	0.009	0.001	0.007
Orbital inclination (deg)	0.05	0.47	0.20	0.21
Crustal composition	Rock	Ice	Ice	Ice
Impact craters	None	Few	Many	Saturated
Surface geology	Volcanic	Tectonic	Tectonic	Static

around Jupiter in prograde orbits that are almost circular (their orbital eccentricities are very low) and almost coplanar (their orbital inclinations are also small: see table 8.1), indicating that they originated in a rotating disk around the young Jupiter, analogous to the Sun's protoplanetary disk[1].

A striking feature of the Jovian system is a dramatic variation in geologic structure and activity from one moon to the next. The innermost of the four, Io, is the most volcanically active body in the solar system, its entire crust resurfaced with igneous rock and sulfur compounds responsible for its yellowish color. It is also the densest, presumably because volcanism has resulted in loss of H_2O and other volatiles to space. Europa, Ganymede and Callisto have predominantly icy surfaces,

---

[1] Jupiter has at least 65 additional moons, all very small (mean radii $\lesssim$125 km), their combined mass amounting to only 0.003% of the total for all moons in the Jovian system.

but with differing morphologies. As discussed in section 5.2.2, the incidence of visible impact craters provides a means of assessing the degree to which a surface has been modified by endogenous processes. A clear trend is evident: whereas Europa has just a few, some areas of Ganymede and almost the entire surface of Callisto are saturated with craters (meaning that any further impacts would overlap existing ones), comparable with the lunar highlands. This implies that the surface of Callisto has been little altered since the final phase of heavy bombardment, whereas Europa and (to a lesser extent) Ganymede have been resurfaced.

These trends arise because of systematic differences in the level of internal heating generated by tidal friction. All four moons are tidally locked in synchronous rotation, i.e. their orbital and spin periods are the same. As they follow their orbital paths they undergo differing degrees of tidal deformation, arising from the slight ellipticities of their orbits and from their gravitational interactions with each other as well as with Jupiter. For the inner moons, the tidal effect is greatly amplified by the occurrence of orbital resonances: the periods of Europa and Ganymede are in 2:1 and 4:1 resonances with that of Io, respectively (see table 8.1), resulting in alignments that occur repeatedly rather than randomly. The degree of tidal heating is therefore greatest for the innermost and least for the outermost of the moons, and so the trend from hyperactive Io to inert Callisto is readily understood.

8.1.1 Europa

The sparsity of impact craters on Europa suggests that it has been resurfaced within the last 10–100 Ma. The icy crust displays little surface elevation, devoid of tall mountains or deep valleys. Instead, it is criss-crossed by a global pattern of shallow striations attributed to tectonic motion within the crustal ice. Some of these features appear to result from convective spreading of the crust as warmer ice rises from below, analogous to divergent tectonic plate boundaries on Earth. Europa also shows evidence of current cryovolcanic activity—eruptive emission of ices from vents and fissures in the crust—discovered in images taken by the Hubble Space Telescope. It is not certain whether resurfacing resulted entirely from tectonic and cryovolcanic activity or whether the crust has experienced periods of partial melting and refreezing. In any case, the surface of Europa is clearly not a habitable environment: it is not only deep-frozen (the current mean global temperature is ~100 K) and devoid of any significant atmosphere but also exposed to lethal doses of high-energy ions emanating from Jupiter's strong magnetic field. From an astrobiological perspective, the focus is on the possibility of life in a subsurface ocean.

The presence of a subsurface ocean is predicted by models for Europa's internal structure and thermal properties. An initial composition of metal, rock and ice is assumed, in proportions that yield the observed mean density. Gravitational segregation of these materials results in a core-mantle-crust structure, as illustrated in figure 8.2. Estimates of the degree of internal heating from tides and from residual decay of radioactive elements allows temperature and pressure to be estimated as a function of depth, and this leads to the prediction of a liquid ocean beneath the icy

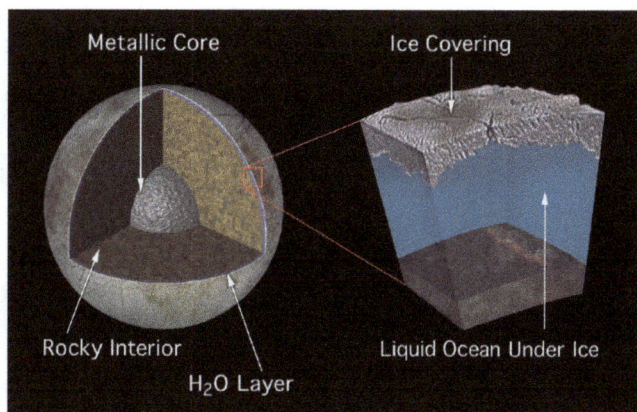

Figure 8.2. Schematic illustrations of the internal structure of Europa, based on a model that predicts the presence of a liquid ocean beneath the icy crust. Image credit: NASA/JPL.

crust. The ice within the crust itself is predicted to vary with depth from a cold, deep-frozen solid at the surface to a warmer, more mobile form below, as it transitions into the liquid phase. The depths of these outer layers are not very tightly constrained by the models. Studies of rare large impact craters on Europa provide some insight, as the projectiles evidently did not fully penetrate the crust, suggesting that it must be at least 10 km thick. Estimates of the thickness of the crust and of the total H_2O content of the body lead to a predicted ocean depth of ~100 km: if this estimate is accurate, Europa has about twice as much liquid water as the Earth (see figure 8.3).

8.1.2 Ganymede

Extensive areas of Ganymede display striated features indicative of tectonic alteration of its icy crust. Unlike Europa, however, resurfacing of Ganymede was not global (areas of heavily-cratered ancient crust still remain), and probably not recent (cratering suggests that the striated regions are at least 500 Ma old). Moreover, no evidence of cryovolcanic activity has been observed on Ganymede. These findings seem consistent with Ganymede's placement in the Jovian system, where it is expected to experience less internal heating than Europa, but considerably more than Callisto.

Models for the structure and thermal properties of Ganymede have been developed, analogous to those for Europa described above. Internal segregation into metal core, rocky mantle and icy outer layers is assumed, but a much larger volume of ice is needed to account for Ganymede's relatively low mean density (see table 8.1). Also, because of its greater mass, the internal pressures are higher, leading to the possibility of phase changes not only between solid and liquid states but also between different crystalline states of solid H_2O. A liquid ocean may be sandwiched between a crust composed on normal (hexagonal) ice and an underlying layer composed of a denser form, as illustrated in figure 8.4. Multilayered models involving additional phase changes are also possible.

Figure 8.3. Jupiter's moons Europa (upper left) and Ganymede (lower left) compared to the Earth, with relative sizes to scale. The blue spheres to the lower right of each image represent the estimated volumes of liquid water contained in the oceans of these bodies, assuming that Europa and Ganymede each have a subsurface ocean of depth 100 km. Image credits: NASA/JPL, augmented by the author.

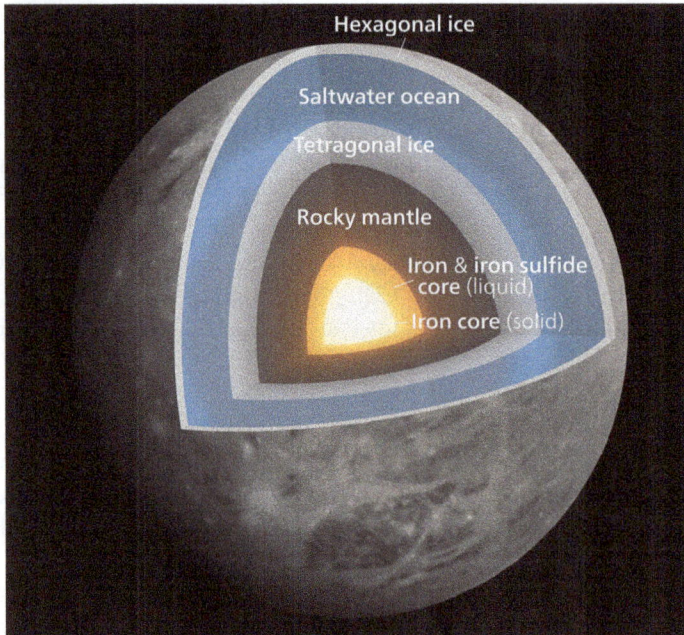

Figure 8.4. Schematic illustration of the internal structure of Ganymede, in which a salty liquid-water ocean is sandwiched between hexagonal and tetragonal crystalline ice phases. The core is thought to contain liquid iron, responsible for Ganymede's intrinsic magnetic field. Image credit: NASA/JPL, edited by the author.

Observational support for the presence of a subsurface ocean is provided by observations of aurorae in the atmosphere of Ganymede. Its atmosphere is extremely tenuous (the surface pressure is a factor of about 10^{11} less than on Earth) and composed primarily of oxygen released from the icy surface by dissociation of H_2O. A weak but measurable aurora is created when atmospheric O_2 molecules are dissociated by collisions with ions from the solar wind that track the ambient magnetic field. Ganymede is immersed in Jupiter's strong magnetic field, but also possesses a magnetic field of its own, thought to be generated by an internal dynamo analogous to that of the Earth. The combination of the two magnetic fields controls the distribution of the aurora, and this changes with time in a predictable way. The observed temporal variations are less than predicted, consistent with the presence of an electrically conductive layer beneath Ganymede's surface, in which an induced magnetic field dampens the variations. The most probable cause is a *saline* subsurface ocean, about 100 km deep, situated beneath a 150 km crust. The volume of water contained by such an ocean is shown schematically relative to Europa and the Earth in figure 8.3.

8.2 Saturn and beyond

Saturn's family of moons differs markedly from that of Jupiter in having only one large member (Titan, radius 2576 km) and several members of moderate size (radii 200–800 km), a range that is unrepresented in the Jovian system. Of these, Enceladus (radius 252 km) is notable as the smallest moon known to have a geologically active surface. Titan is also of great interest as the only moon in the solar system to have a dense atmosphere.

8.2.1 Enceladus

Enceladus (figure 8.5) is one of several inner moons of the Saturnian system, its mean orbital radius corresponding to just over four times the mean radius of Saturn itself. Enceladus is tidally locked, with synchronized orbital and spin periods of 1.37 days, and is in a 1:2 orbital resonance with a more distant and more massive moon (Dione). The remarkable degree of surface activity for so small a world is attributed to the amplification of tidal heating within Enceladus by this resonance. Features produced by tectonic alteration of the crust include cracks and grooves, extending over much of the surface, most prominent in the south-polar region. Impact craters persist in some areas, many of them degraded by viscous motion and relaxation of the crustal ice. Models for the interior of Enceladus suggest the presence of a global ocean of mean depth about 25–30 km, beneath a 30–40 km-thick crust.

Enceladus displays frequent cryovolcanic activity. The most active area is near the south pole, where a series of grooves known as the 'tiger stripes' regularly vent plumes of ice, dust and gas (figure 8.6). Many of the particles reach escape velocity (0.24 km s^{-1}), feeding into a diffuse ring (the E-ring) around Saturn, beyond the planet's main ring system. A chemical analysis of the plume material carried out by the Cassini–Huygens mission showed that it contains CO, CO_2, CH_4 and organic molecules in addition to H_2O, in proportions similar to those observed in comets

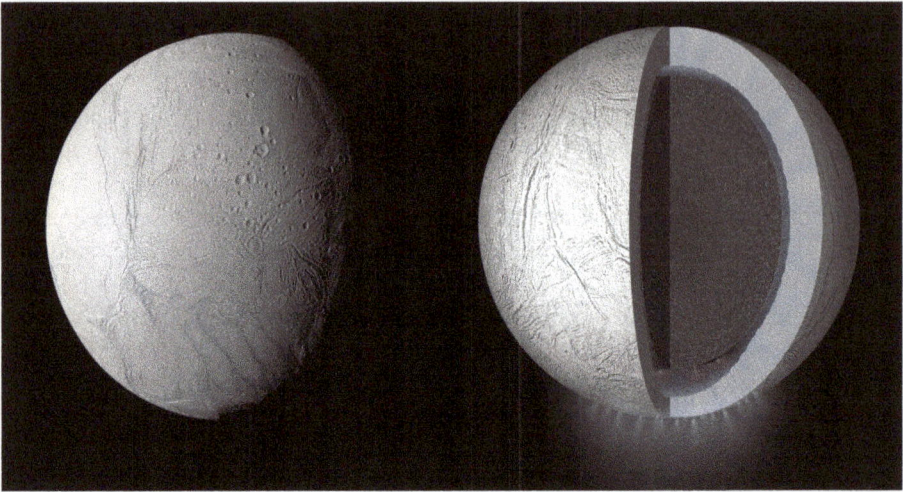

Figure 8.5. Enceladus. Left: color-enhanced mosaic of images obtained by the Cassini spacecraft, illustrating variations in surface terrain. Right: a schematic model for the internal structure, including a rocky core (dark grey), liquid ocean (dark blue) and icy crust (light blue). A hydrothermal origin for cryovolcanic activity near the south pole is suggested. Image credit: NASA/JPL/Cassini-Huygens Mission.

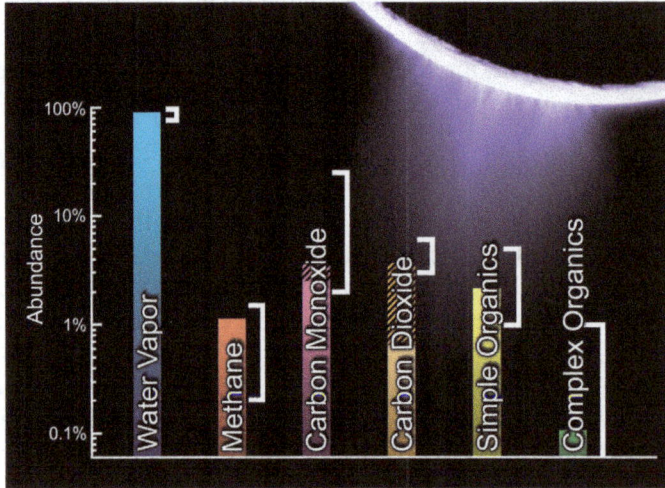

Figure 8.6. Chemical analysis of gases vented from the surface of Enceladus, based on data collected by the Cassini spacecraft as it flew through a cryovolcanic plume (shown in a color-enhanced image, top right). Note the logarithmic abundance scale. The white bracket beside each column indicates the range of values observed in comets. Image credit: NASA/JPL/Cassini-Huygens Mission.

(figure 8.6), as expected for an icy world that accreted from comet-like planetesimals. The D/H ratio of the ice is also typical of cometary values (see figure 5.6). Molecular hydrogen has also been detected.

8.2.2 Titan

Titan contains 96% of the mass of all material (moons and ring particles) in orbit around Saturn. Because of this, tidal heating is effectively independent of other moons in the system and arises only through Titan's gravitational interaction with Saturn itself. Titan maintains a respectable distance from Saturn—the mean orbital radius (1.22×10^6 km) is greater than that of Ganymede in the Jovian system—but the orbital eccentricity (0.029) is appreciably higher than for any of the Galilean moons, and a modest degree of tidal flexing is thus predicted. Detailed measurements by the Cassini spacecraft indicate a degree of tidal deformation greater than expected for a solid body but consistent with the presence of a 100 km-thick crust floating on a global subsurface ocean (figure 8.7). Mixing of NH_3 with H_2O, which lowers the melting point, may help to maintain the ocean in a liquid state. In common with models for Ganymede (figure 8.4), Titan's ocean is predicted to overlie a denser internal ice layer that separates it from the rocky core.

Perhaps the most remarkable features of Titan are its atmosphere and resultant surface conditions. The atmosphere is composed of N_2 (95%) and CH_4 (4.9%) by volume, together with smaller concentrations of other hydrocarbons and CN compounds. The low surface temperature (~95 K) ensures that any CO_2 and H_2O would freeze out of the atmosphere. The surface pressure is a factor of 1.5 greater than on Earth. That Titan should retain so dense an atmosphere is not necessarily surprising, as N_2 is expected to be reasonably stable with respect to thermal leakage into space (indeed, it is perhaps more remarkable that Ganymede lacks a dense atmosphere; see figure 4.4). Because of its lower gravity, Titan's atmosphere extends to greater heights compared to the Earth's. In the upper layers, CH_4 is subject to photodissociation by solar UV radiation, generating free H and radicals that interact to form more complex hydrocarbons. This results in a thick orange haze that renders

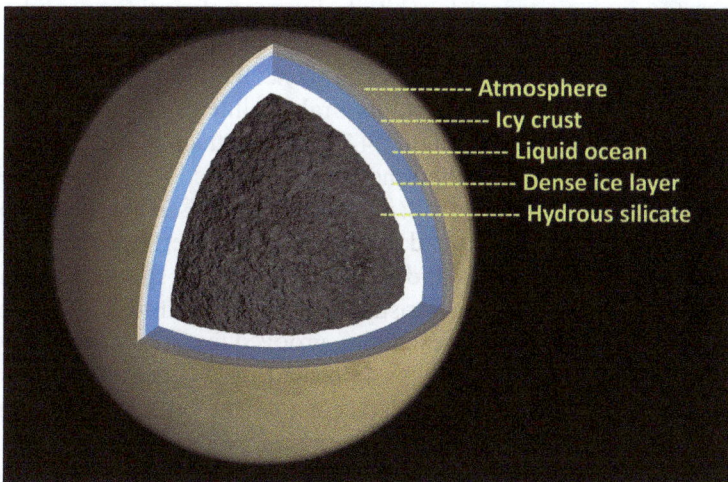

Figure 8.7. Schematic illustration of a model for the internal structure of Titan, in which a liquid H_2O–NH_3 ocean is sandwiched between the icy crust and denser crystalline ice, layered upon a hydrous silicate core. Image credit: NASA/JPL, annotated by the author.

Titan's atmosphere almost opaque to visible light: only ∼10% of incident sunlight reaches the surface. The timescale for consumption of all of Titan's atmospheric CH_4 in this way is estimated to be about 30 Ma. Its continued presence thus implies that it is being replenished, perhaps by cryovolcanic outgasing.

The temperatures and pressures prevailing on the surface of Titan are such that CH_4 and other hydrocarbons may transition to and from the liquid phase, in a similar manner to the behavior of H_2O on Earth. The presence of hydrocarbon lakes was confirmed by data collected by the Cassini–Huygens mission, and these appear to be fed by a system of river-like dendritic channels, suggestive of a cycle of precipitation, flow and evaporation, analogous to the terrestrial hydrological cycle. This discovery has led to speculation that an exotic form of life might be possible on Titan, in which CH_4 substitutes for H_2O as the primary fluid: see section 8.3.3 for further discussion.

Looking to the future, when the Sun enters its red giant phase some 4–5 Ga from now, temperatures are predicted to become warm enough for liquid water to flow on the surface of Titan. By then, the Sun's luminosity will have increased by a factor of approximately 100, and the circumstellar habitable zone (section 4.4.1) will have migrated outward to encompass the Saturnian system. Titan might thus become a candidate to host life as we know it, perhaps arising from an indigenous origin, or perhaps as a refuge for fleeing earthlings.

8.2.3 Triton, Pluto and Charon

The Uranian system lacks any large moons or strong candidates for subsurface oceans. Moving on to Neptune, however, we find a more promising case. Triton (mean radius 1353 km, about 13% smaller than Europa) is the only moon of significant size in the Neptunian system, and is unique in being the only large moon in the solar system with a retrograde orbit: its direction of motion around Neptune is clockwise from north, opposite to that of Neptune's spin, and its orbit is tilted at an angle of 23° relative to Neptune's equatorial plane. This configuration implies that Triton is a captured body, rather than a natural satellite of Neptune, and may have originated in the Kuiper belt (see figure 4.2). Larger than Pluto, Triton would be considered a dwarf planet if it orbited the Sun. The direction of motion results in a slow and steady decay of Triton's orbit: in contrast to the Earth's Moon (section 5.1) it is gradually spiralling inward instead of outward, until (perhaps 3–4 Ga from now) it will be torn apart by tidal forces and Neptune will gain a spectacular ring system.

Triton's surface displays characteristics qualitatively similar to those seen on Europa and Enceladus: cryovolcanic activity, grooved features attributed to tectonic motion of the icy crust, and resurfacing indicated by a general sparsity of impact craters. The heat driving this activity is accounted for by a combination of tidal friction arising from its gravitational interaction with Neptune and decay of radioactive elements within Triton. Its internal structure may resemble that of Titan (figure 8.7), with a mantle composed largely of H_2O layered between a rocky core and an icy crust. Triton's mantle is not expected to contain high-density ice phases, but may include a liquid ocean.

Finally, Pluto and its moon Charon (radii 1190 and 606 km) form a binary pair of icy worlds on the outer fringes of the planetary system. In 2015 the New Horizons space probe became the first to visit them. At such large solar distances, surface temperatures (40–55 K) are low enough to freeze N_2, which is the primary constituent of Pluto's surface. Both Pluto and Charon show variations in impact crater counts and surface morphology, indicating that partial resurfacing has occurred. Each is presumed to be internally differentiated into a rocky core and an icy mantle and crust, but it is not known whether internal heating is (or ever was) sufficient to permit liquid water to be present. Resurfacing may be explained, at least in part, by the energy delivered by large impacts.

8.3 Prospects for life

As noted at the outset of this chapter, the existence of potentially habitable environments provides no guarantee that they contain life. This section assesses the prospects for life in icy worlds, and possible means of detection.

8.3.1 Hydrothermal systems

The rock–water interface is considered a likely setting for the origin of life on Earth, as reviewed in chapter 6, and this is a natural focus for discussion of an origin of life on icy worlds as well. A world with a global ocean in direct contact with a rocky interior is exceptionally well endowed in this regard. Models for Europa and Enceladus (figure 8.2 and 8.5) predict this scenario, and Triton is also a candidate. Models for the more massive moons Ganymede and Titan (figure 8.4 and 8.7) predict the presence of a dense ice phase beneath the ocean, separating it from the rocky interior. This does not necessarily preclude the possibility of life in their oceans, but it renders comparisons with the Earth less pertinent.

Do the ocean floors of moons such as Europa and Enceladus contain hydrothermal systems analogous to those on Earth? We have no direct means of detecting them but circumstantial evidence is accumulating to suggest that they do. Heat generated in the core and mantle must gradually diffuse outward through the rock–water interface to the ocean, as it does on Earth. Variations in surface morphology suggest that heating is not uniform but may tend to concentrate in certain locations, especially on Enceladus, where the most active features are in the south polar region (figure 8.5). It is logical to associate activity at the surface with corresponding activity on the ocean floor below. The Cassini spacecraft provided support for this interpretation when it sampled cryovolcanic emissions from Enceladus: the plumes were found to contain significant concentrations of gaseous H_2 as well as H_2O, CH_4, NH_3, CO, CO_2 and organic molecules (figure 8.6). Of these, H_2 is exceptional because it requires an active source; the other species are expected constituents of a body assembled from icy, comet-like planetesimals. H_2 clearly emanates from the plumes and cannot be explained by (much lower) levels arising from photodissociation of surface H_2O ice. The authors of the research (Waite *et al* 2017) conclude that the most plausible source of H_2 in the plumes is ongoing geochemical activity, emulating processes that occur in hydrothermal systems on Earth, such as serpentinization (section 5.4.2) and iron–sulfur chemistry (section 6.1.3).

In summary, if hydrothermal systems are, indeed, implicated in the origin of terrestrial life, it is plausible to suppose that life could arise in similar fashion on these icy worlds. The essential ingredients—water, organic molecules and an energy source—are present. Hotspots associated with cracks and vents in the rock layer beneath the ocean may generate dynamic circulation of hydrothermal fluids, to provide a continuous source of reactants and a range of physical conditions for reactions to occur, as is the case on Earth.

8.3.2 Subglacial lakes as analog environments

Subglacial lakes represent another terrestrial phenomenon that informs our search for life on icy worlds. Over 400 have been found under the Antarctic ice sheet, of which the largest is Lake Vostok (figure 8.8). These freshwater lakes, described by Siegert *et al* (2015) as 'one of the planet's last natural frontiers', are of intrinsic interest to biologists as potential habitats for terrestrial organisms that may have survived in isolation from the global biosphere for 10 thousand years or more. In the astrobiological context, they are important for both scientific and technical reasons: they provide an opportunity to enhance our understanding of microbial survival and growth under relevant conditions, and the techniques developed to study them on

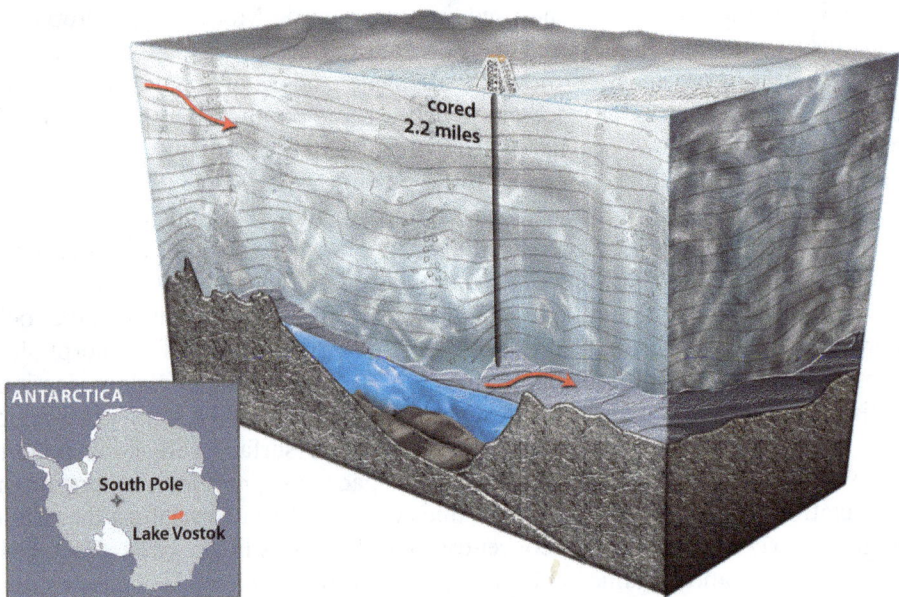

Figure 8.8. Schematic cross-section of Lake Vostok, Antarctica, the largest known terrestrial subglacial lake, situated beneath a 4 km-thick ice sheet. The geographic location is shown in the inset. Red arrows indicate the direction of motion of the ice sheet. Lake Vostok is 250 km long, up to 50 km wide, and has a mean depth of about 0.5 km. It contains an estimated volume of water somewhat greater than Lake Michigan. Core drilling is used to obtain samples for analysis. Image credit: Nicolle Rager-Fuller/NSF.

Earth may also be applicable to future investigations of such environments on other bodies in our planetary system.

Subglacial lakes were first detected by radar sounding techniques, using instruments carried by aircraft and orbiting satellites. Radio waves reflected from the various possible interfaces (air–ice, ice–water, water–sediment/bedrock and ice–bedrock) are analyzed to map the distribution, depths and dimensions of the lakes. The topography of the bedrock beneath the ice sheet was found to be conducive to lake formation, the lakes occupying topographic valleys and basins. Geothermal heat diffusion warms the water from below, and the overlying ice sheet insulates it from freezing temperatures on the surface. High ambient pressures reduce the melting point of the lake water to about −3 C. The ice sheet is thought to have formed some 15 Ma years ago, but the lakes have not remained in stable isolation since then. The sheet is in continuous motion (figure 8.8), and this results in gradual replenishment: water in contact with the drifting ice sheet freezes as 'accretion ice' and is carried away, to be replaced by melting ice in warmer regions. The estimated timescale to recycle the entire volume of a large lake is 10 000–15 000 years. Also, in many cases, the lakes are not isolated from each other but are interconnected by a network of subglacial rivers and streams.

Samples have been collected and analyzed from Lake Vostok, and also from another subglacial Antarctic lake (Lake Whillans), a much smaller, shallower and more accessible target. To accomplish this, a borehole is drilled into the ice sheet, and samples are collected from different depths. Figure 8.8 shows drilling to a depth just short of the surface of Lake Vostok, a point at which microbial life from the lake may be encapsulated in the accretion ice. Drilling all the way into the lake was attempted for the first time in 2012. Hydraulic pressure caused water to flow into the borehole, where it froze, enabling solidified samples of lake water to be recovered. A similar procedure was used to collect samples from Lake Whillans, but with an important difference. A key problem is to maintain the mobility of the drill at the high pressures and low temperatures encountered in the ice sheet. Drilling into Lake Vostok (through 4 km of ice!) necessitated a mechanical drill lubricated with kerosene; at Lake Whillans, where the overlying ice is only 0.8 km thick, it was possible to use a hot-water drill that did not require a lubricant, thus avoiding an obvious source of contamination.

Analysis of samples from Lake Whillans revealed the presence of a diverse microbial community composed of chemosynthetic bacteria and archaea, providing unequivocal evidence that this subglacial environment hosts an active ecosystem that flourishes in the absence of sunlight (Mikucki *et al* 2016). However, the geological setting and small volume of this lake imply a recycling timescale of only a decade or so, a factor of ~10^3 less than estimates for Lake Vostok, so this finding does not address the question of whether such communities can survive in isolation for much longer periods. Results so far obtained from Lake Vostok are affected by problems of contamination by surface biota as well as by drilling chemicals. After exclusion of probable contaminants, a few apparently indigenous bacterial species have been reported in both the accretion ice and the refrozen lake-water samples, but the overall biomass seems to be small.

8.3.3 Alternatives to water?

The preceding discussion is based on the assumption that liquid water is a requirement for life, and this limits the possibilities on icy worlds to subsurface environments. However, the discovery that liquid methane is stable on the surface of Titan prompts an assessment of methane or other potential alternatives to water as a fluid medium for biological processes. The salient properties of water, and the vital roles it performs in terrestrial biology, would presumably need to be proxied by an alternative fluid. These are summarized below.

- H_2O is a *polar* molecule (the distribution of charge is asymmetric, so it acts as a dipole), and also *amphoteric* (it can either donate or accept an H^+ ion, allowing it to act as an acid or a base). It is thus an excellent solvent for polar and ionic compounds, and ideally suited to its role as the medium for biochemical reactions.
- Water plays vital roles in regulating biomolecular structure and functionality, by means of self-organizing processes imposed by hydrophobic effects. These include compartmentalization into cells, protein folding, enzyme formation and function, transcription and reproduction.
- Water has a high heat capacity and is less dense in solid phase than in liquid phase. These properties help to regulate temperature and enable liquid to persist beneath surface ice, whereas most other fluids freeze solid at the equivalent phase boundary.

Water appears to be the only abundant fluid that satisfies all of these criteria (see Pohorille and Pratt 2012 for a review). Formamide (CH_3NO) is perhaps the most feasible alternative—it is polar, like H_2O, and capable of fulfilling several of its biological roles—but, also like water, it would be frozen solid on the surface of Titan.

In contrast, CH_4 and other liquid hydrocarbons in Titan's lakes are non-polar. This does not necessarily rule them out as solvents enabling chemical reactions analogous to some of those occurring in terrestrial biology; but the possibilities are more limited, and organic molecules with alternative functional groups and alternative structural frameworks would be required. For example, vinyl cyanide (C_2H_3CN), a species detected in Titan's atmosphere, has been suggested as a possible alternative to phospholipids as a molecule capable of forming membranes in hydrocarbon fluids. Reduction of higher-molecular-weight hydrocarbons such as acetylene (C_2H_2) and ethane (C_2H_6) to CH_4 has been proposed as an alternative form of methanogenic metabolism that might be possible on Titan (McKay and Smith 2005).

8.3.4 Future exploration

Missions designed to search for life in subsurface oceans on icy worlds might adopt either of two strategies. The most ambitious plan would be to deliver a landing vehicle with the capability to drill through the crust to the ocean and collect samples for analysis—an approach analogous with those currently being used to explore subglacial lakes on Earth (section 8.3.2). However, the technology to accomplish this is still far in the future: we have not yet perfected techniques to collect uncontaminated samples from beneath a 4 km ice layer on Earth, far less from beneath an icy crust >10 km thick on a distant world in the outer solar system. A much more realistic plan is to develop new and more sophisticated instruments for non-invasive exploration of these worlds, and to collect samples that they might serendipitously deliver to us.

The samples collected by the Cassini spacecraft from the cryovolcanic plumes of Enceladus (section 8.2.1) present a clear demonstration of principle for the latter approach. Evidence is accumulating that Europa is also cryovolcanically active, motivating plans for sample collection and analysis by a future mission to this moon. Sensitive instruments may be used to determine the chemical composition of the plumes and to search for potential biomolecules such as amino acids and nucleotides. The ultimate goal would be to test for the presence of microbial cells. Further data are also needed to enhance our understanding of geological processes on icy worlds, including the degree of interaction and circulation of materials between crustal and subsurface layers, and to place tighter constraints on models for internal structure.

Missions under development to address such questions include the Jupiter Icy Moons Explorer (JUICE, a European Space Agency mission) and NASA's Europa Clipper mission, both scheduled for launch in or around 2022. Both will collect data during multiple flybys (neither will carry a lander vehicle). JUICE will study all three icy Galilean moons of Jupiter (Europa, Ganymede and Callisto) to enable detailed comparisons of their geology, structure and surface composition. The Europa Clipper mission will focus on assessing the habitability of Europa. It will use infrared imaging to identify potentially active regions, and spectrometers to analyse the composition of both surface and plume materials. It will also attempt to identify the most appropriate landing site for a future lander mission to Europa.

Questions and discussion topics

- Why do orbital resonances between members of a moon system affect tidal heating?
- Explain how detailed study of the external properties of an icy world provides insight into internal structure and subsurface processes.
- How can the presence of hydrothermal systems in the subsurface ocean of an icy world be inferred without direct observation?
- With reference to information in section 4.4.1, verify that a 100-fold increase in solar luminosity, predicted to occur when the Sun becomes a red giant, will place the Saturnian system in the circumstellar habitable zone.
- Explain why the direction of Triton's orbital motion causes it to slowly spiral inward toward Neptune.
- Suppose you are in charge of a major space agency with funds to support a future sample return mission to either Mars or Europa, but not both: which would you choose, and why? Assume that either mission would have deep drilling capabilities.

References and further reading

Bulat S A 2016 Microbiology of the subglacial Lake Vostok: First results of borehole-frozen lake water analysis and prospects for searching for lake inhabitants *Phil. Trans. R. Soc.* A **374** 20140292

Lorenz R D 2016 Europa ocean sampling by plume flythrough: Astrobiological expectations *Icarus* **267** 217

McKay C P and Smith H D 2005 Possibilities for methanogenic life in liquid methane on the surface of Titan *Icarus* **1778** 274

Mikucki J A *et al* 2016 Subglacial Lake Whillans microbial biogeochemistry: a synthesis of current knowledge *Phil. Trans. R. Soc.* A **374** 20140290

Pohorille A and Pratt L R 2012 Is water the universal solvent for life? *Orig. Life Evol. Biosph.* **42** 405

Siegert M J, Priscu J C, Alekhina I A, Wadham J L and Lyons W B 2015 Antarctic subglacial lake exploration: first results and future plans *Phil. Trans. R. Soc.* A **374** 20140466

Sparks W B *et al* 2017 Active cryovolcanism on Europa? *Astrophys. J.* **839** L18

Spencer J R and Nimmo F 2013 Enceladus: An active ice world in the Saturn system *Ann. Rev. Earth Planet. Sci.* **41** 693

Tobie G *et al* 2014 Science goals and mission concept for the future exploration of Titan and Enceladus *Planet. Space Sci.* **104** 59

Waite J H *et al* 2017 Cassini finds molecular hydrogen in the Enceladus plume: Evidence for hydrothermal processes *Science* **356** 155

IOP Concise Physics

Origins of Life
A cosmic perspective
Douglas Whittet

Chapter 9

The search for life beyond our solar system

Identification of potentially habitable worlds beyond our solar system has become a realistic endeavour for the first time in human history. Elegant and exquisitely precise techniques have been developed to detect and characterize planets orbiting other stars, and examples have been found that appear to be similar to the Earth in size and global temperature. The next step is to search for unambiguous signatures of life on such planets. Whether life will be found remains an open question, of course, because we do not yet know whether life is an inevitable outcome of the appropriate conditions or an extremely improbable occurrence under any circumstances. However, our technology is advancing to a point where we can predict with some confidence that if life (as we know it) is widespread in our Galaxy, we will surely find it, perhaps sooner rather than later. A confirmed detection would answer one of the most fundamental questions in science. In contrast, non-detections do not answer it because absence of evidence is not evidence of absence, in which case the quest will continue.

This chapter reviews what we have learned about the properties of exoplanets, their potential for life, and how best to search for it. Both spectroscopic biosignatures and SETI are considered. The final section suggests some possible implications of a successful outcome.

9.1 Exoplanetary systems: characteristics and habitability

9.1.1 Detection

The first confirmed detection of an exoplanet orbiting a main-sequence star was made in 1995[1]. Since then, the number of detections has risen at an exponential rate as observational facilities and detection methods have been developed and refined, culminating in the 2009 launch of the Kepler mission, a space observatory dedicated

---

[1] Planets orbiting a pulsar—the stellar remnant of supernova explosion—were found in 1992.

doi:10.1088/978-1-6817-4676-0ch9

to exoplanet detection. With very few exceptions, exoplanets cannot be imaged directly, because of the overwhelming brightness of the parent star relative to the planet and the very small angular separation between them as viewed from Earth. A range of indirect techniques is used, of which by far the most productive are the transit method and the Doppler (radial velocity) method. Both are, in effect, extreme applications of tried and tested techniques used by astronomers for many years to detect and study binary and multiple star systems.

A detailed discussion of the observational techniques used to detect exoplanets is beyond the scope of this book (see Wright and Gaudi (2013) for a comprehensive review). A summary of the primary methods is given here as a guide to what can be learned from them.

- *Transit method:* If the orbital path of a planet causes it to pass between us and its parent star, a small dip in intensity is observed (figure 9.1). This method is responsible for the large majority of exoplanet detections, including those by the Kepler mission. An obvious limitation is that only planets with orbital planes viewed approximately edge-on are detected, so statistical arguments must be used to estimate the frequency of all exoplanetary systems from the number detected by this method. Transits must also be distinguished from other possible causes of variation in the light from the star, such as star spots (analogous to sunspots) that may appear and disappear as the star rotates. It is essential to observe several successive transits to confirm a detection and determine an orbital period. The amplitude of the dip in intensity provides a measure of the cross-sectional area, and hence the radius, of the planet.

- *Doppler method:* Each planet in a given system orbits around its common center of mass with the host star. In general, the planetary mass is very small compared with the stellar mass, so the center of mass is close to the center of the star; but with sufficient sensitivity, the resultant motion of the star may be observed as a cyclic Doppler shift in its spectrum (figure 9.2). With the exception of orbital planes viewed face-on (perpendicular to the line of sight), any orientation will produce a signal. The presence of multiple planets in a

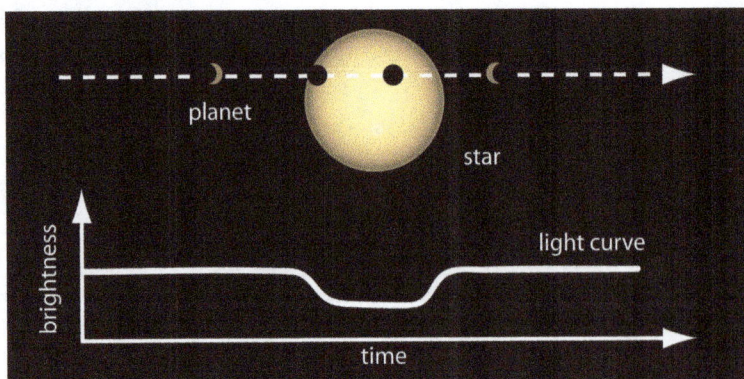

Figure 9.1. Schematic illustration of the transit method of exoplanet detection. A dip in brightness is observed as the planet passes in front of the star. Image credit: NASA/ARC.

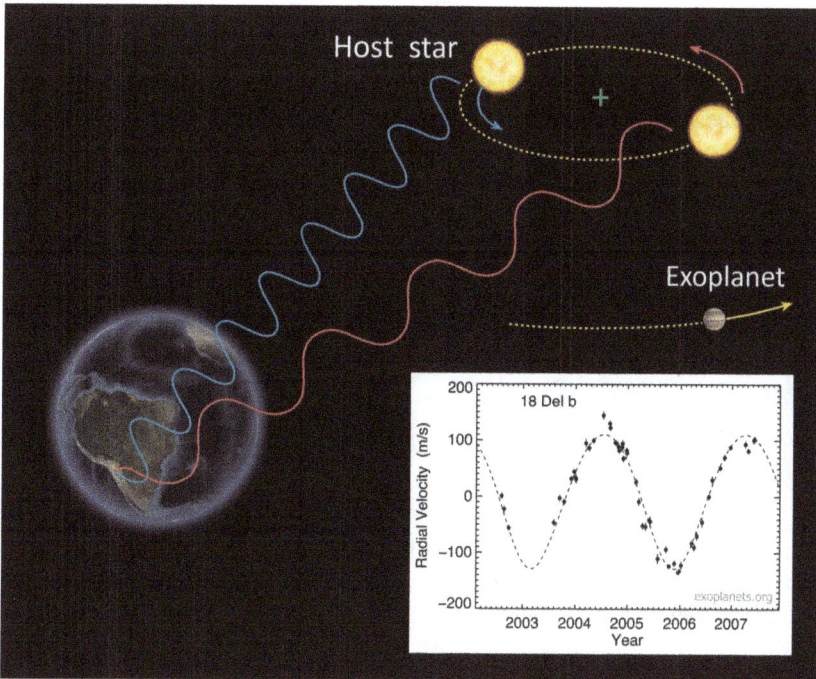

Figure 9.2. Schematic illustration of the Doppler method of exoplanet detection. Host star and exoplanet are in orbit about their common center of mass (green cross), and the resultant periodic motion of the star (greatly exaggerated in this illustration!) is observed as radial velocity variations calculated from the Doppler shift in its spectrum. The inset shows an example. Image credit (schematic): European Southern Observatory, annotated by the author; inset image based on data from the Exoplanet Orbit Database and Exoplanet Data Explorer at exoplanets.org (Han *et al* 2014).

system results in overlapping variations with different periods that must be disentangled to enable individual planets to be studied.

The two methods complement each other. Combining data from both types of detection is ideal as it enables the characteristics of an exoplanet to be fully explored. Detections by the transit method are therefore followed by spectroscopic observations of the same stars. In principle, the velocity data enable the characteristics of the planet's orbit (period, semi-major axis, eccentricity) and also the mass of the planet to be determined, using conservation of momentum and Kepler's 3rd law of planetary motion. In general, however, only a *minimum* value for the planetary mass ($m \sin i$, where m is the actual mass) can be measured in this way, as i (the inclination of the orbital plane relative to the perpendicular to the line of sight) is usually unknown. But in a transiting system we know that i must be close to 90°. The observations then provide an estimate of the actual mass of the planet, together with its radius and hence density. The density, in turn, carries information on composition, distinguishing, for example, a planet composed mostly of rock and metals (3000–6000 kg m^{-3}) from one composed mostly of rock and solid or liquid H_2O (1500–3000 kg m^{-3}).

It is important to note that additional selection effects influence detection. The strength of the 'signal' for each method (the depth of the transit dip and the amplitude of the Doppler variation) depends on the size and mass of the planet, hence larger ones are far more readily detected than smaller ones. Planets that orbit close to the star are also easier to detect than those in more distant orbits because the signal is relatively strong and the period is relatively short, so many cycles may be observed in a reasonable time. A planet with an orbital period exceeding that of Saturn (29 years), for example, has yet to complete a single orbit since the first exoplanet detections were made. Progress toward full sampling of exoplanetary systems is thus a long-term project. The following sections review what has been learned to date.

9.1.2 Metallicities of host stars

A supply of 'heavy' elements is needed to form planets. This is most obvious in the case of rocky planets, composed primarily of elements such as Mg, Si, Fe and O, but also true of gas giants if they are seeded by rocky/metallic embryos (section 4.2.4). How long did it take for the build-up of heavy elements (section 2.3.3) to be sufficient for planet formation to be triggered? Evidence that addresses this question is shown in figure 9.3, which plots a histogram of exoplanet detections according to the metallicity index of the parent star. Based on spectroscopic analysis of the star's atmosphere, the metallicity index is closely correlated with heavy-element availability in the protoplanetary disk of the system. The peak of the distribution corresponds quite closely to the solar metallicity—our home star was well endowed with resources for planet parenthood! In contrast, very few stars in the sample have an index below −0.6, corresponding to 25% of solar metallicity. Comparing figure

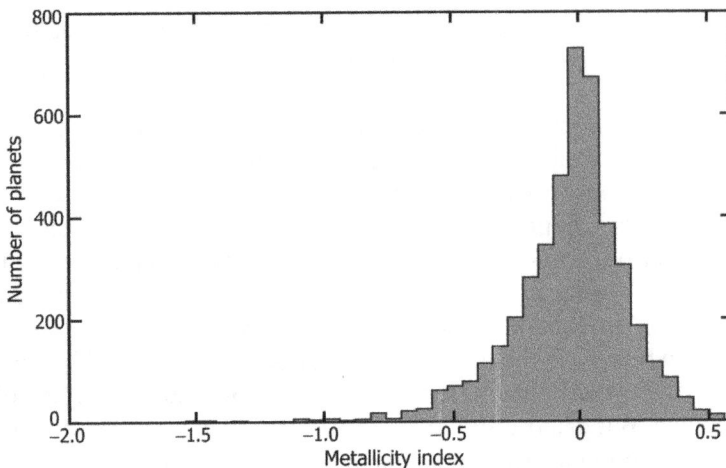

Figure 9.3. Histogram showing the distribution of detected exoplanets with regard to the metallicity index of the parent star. Metallicity indices of −2, −1 and 0 correspond to 1%, 10% and 100% of the solar metallic-element abundances, respectively. Based on data from the Exoplanet Orbit Database and Exoplanet Data Explorer at exoplanets.org (Han *et al* 2014).

9.3 with figure 2.6, it appears that some 2–4 Ga of evolution may have been needed to deliver this level of enrichment, following the birth of the Galaxy from primordial H and He some 14 Ga ago. This estimate is consistent with observations of the oldest stars known to have Earth-sized planets, which are about 11 Ga old (Campante *et al* 2015), i.e. they formed when the Galaxy was about 20% of its current age—perhaps a threshold for rocky planet formation.

9.1.3 Circumstellar habitable zones

A simple rule of thumb for the size of the circumstellar habitable zone (section 4.4.1) around a star of luminosity L is that its inner and outer radii vary in proportion to $\sqrt{L}$. The habitable zone around a star of lesser or greater luminosity relative to the Sun is correspondingly closer in or further out compared with the solar system's habitable zone. This is illustrated in figure 9.4 for stars in the range 0.25–4.0 times the solar luminosity. Additional factors come into play if this range is extended further. The main-sequence lifetime of a star (corresponding to the stable hydrogen-burning phase; section 2.2) varies inversely with luminosity, and is <3.6 Ga for a star of luminosity >4 times solar: if our Sun had been this luminous, it would already have evolved to become a red giant. Consequently, stars much more luminous than the Sun are not considered viable long-term hosts for habitable planets. In any case, such stars represent only a small fraction (~3%) of the total main-sequence population. In contrast, stars less luminous than the Sun have life expectancies >10 Ga and are also very common (~90% of the population).

Low-luminosity (red dwarf) stars present different challenges as potential hosts for life. They are not only less luminous that the Sun but also cooler (surface

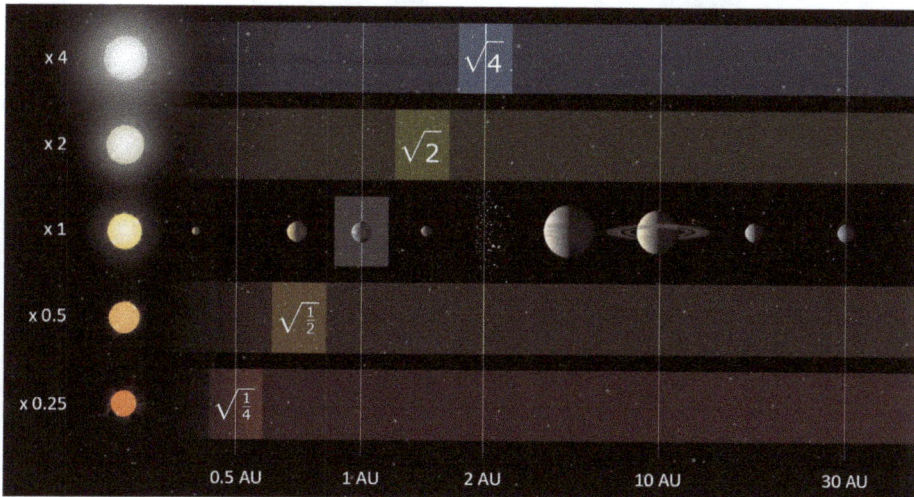

Figure 9.4. Schematic illustration of the distance–luminosity relation for circumstellar habitable zones ($r \propto \sqrt{L}$; section 4.4.1), comparing the Sun with other main-sequence stars. Vertical and horizontal scales indicate luminosity (relative to solar) and radial distance, respectively. Image adapted by the author from an original by Derpedde (Wikimedia Commons).

temperatures in the range 2300–4500 K). Because of this they emit most energy at red and infrared wavelengths, as illustrated in figure 9.5: Planck (blackbody) curves of appropriate temperature represent the variation in brightness with wavelength in each case. Light-dependent biological processes such as photosynthesis (section 9.2.3) may be affected. Another difference is that red dwarfs are typically more active than Sun-like stars, displaying transient flare activity analogous to solar flares, but of greater intensity and frequency, often resulting in large changes in luminosity as well as bursts of ionizing radiation and streams of energetic particles.

Being correspondingly closer to the star, a planet in the habitable zone of a red dwarf is liable to become tidally locked. A planet locked into synchronous rotation in a circular orbit with low obliquity would experience constant daylight on the star-facing side and constant night-time on the opposite side. In our solar system, only planets inside the inner radius of the habitable zone are affected, and Mercury has become tidally locked into a 3:2 spin-orbit resonance rather than a synchronous (1:1) resonance. The probability of synchronous resonance for a planet in the habitable zone increases for less luminous stars, and becomes almost inevitable for those with luminosities <10% of solar. Synchronous rotation does not necessarily rule out the possibility of life on a planet, but it would surely be challenging: temperatures cold enough to freeze atmospheric gases and create a permanent hemispherical ice cap are likely to occur on the side facing away from the star.

So far we have considered only single stars, but what of binary and multiple systems? Their occurrence in the stellar population as a whole is not precisely known but is probably at least 30%–50%. In contrast, approximately 5% of confirmed exoplanet detections occur in binary star systems, and a further 1% in systems with more than two stars, so it appears that such systems are less likely to host planets than single stars. This is not unexpected, as situations that allow a planet to survive in a stable orbit are generally less probable when two or more stars are involved.

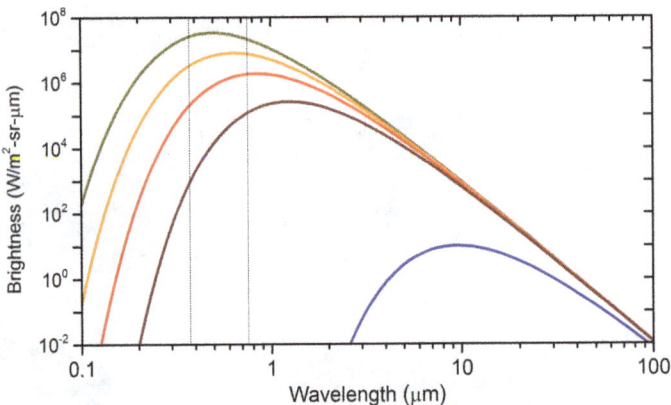

Figure 9.5. Comparison of Planck curves representing the continuous spectra of bodies of different surface temperatures: the Sun (5778 K, yellow) and three examples spanning the temperature range of red dwarfs (4500 K, orange; 3400 K, red; 2300 K, wine). Also shown is a 300 K curve (blue), representative of the thermal infrared emission of a planet in the circumstellar habitable zone of a star. Vertical dotted lines indicate the range of the visible spectrum.

Two scenarios that do offer prospects of long-term orbital stability under potentially habitable conditions in a binary system are summarized below.

- A circumbinary planet orbits about the center of mass of a relatively close pair of stars at a radial distance much larger than their separation. An inhabitant of such a planet would see two 'suns' instead of one, always close together in the sky. The sum of their luminosities determines the locus of the habitable zone, with only minor deviations from that of a single star of equivalent total luminosity.
- A planet orbits one member of a wide binary system at a radial distance much smaller than the separation of the two stars. In this case, the nearby star would generally be the primary source of radiation (assuming it is not significantly less luminous than the other), so the locus of the habitable zone would not be greatly affected by the presence of the companion star.

9.1.4 Exoplanetary systems compared

An informative overview of the exoplanet population is provided by the distribution of size with respect to orbital period, shown in figure 9.6. The orbital period increases with distance from the star according to Kepler's 3rd law, so planets to the left of the plot are generally much closer to their host star than any planet in our solar system is to the Sun. Various groupings are illustrated, and they include

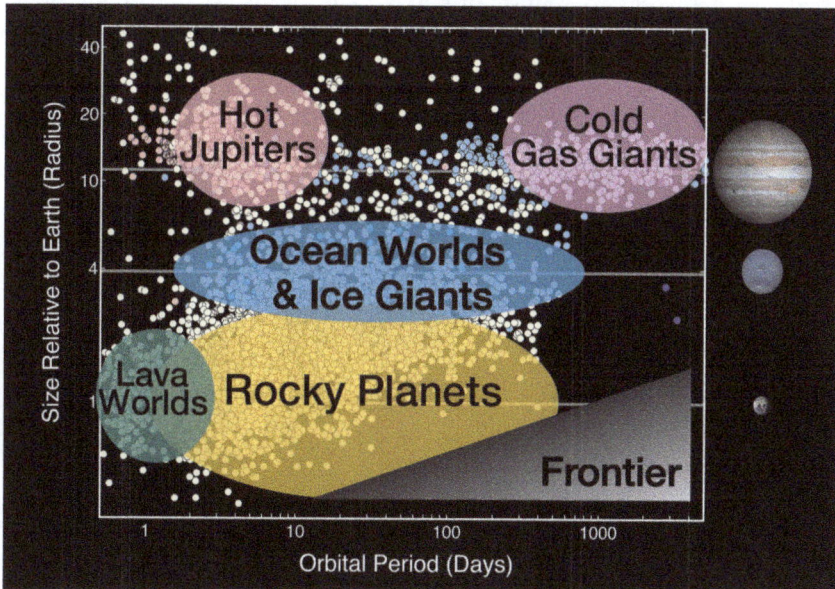

Figure 9.6. Plot of exoplanet size versus orbital period. Each small circle represents a planet, color-coded by detection method: transits observed by the Kepler spacecraft (cream); transits observed by other telescopes (pink); Doppler method (pale blue); gravitational lensing (dark blue). Colored ellipses indicate major groupings, and the area labeled 'Frontier' is yet to be explored. The Earth, Neptune and Jupiter (images and horizontal lines) indicate representative planetary sizes in our solar system. Image credits: NASA/ARC, Natalie Batalha and Wendy Stenzel.

planetary types that differ quite markedly from our familiar local ones. Hot planets are evidently quite common, including giant 'hot Jupiters' and terrestrial-sized 'lava worlds'. Planets intermediate in size between the Earth and Neptune, a range unrepresented in our solar system, are very common. Neptune-sized planets may be 'ice giants' (like Neptune itself) or 'ocean worlds', depending on their location, whereas Earth-sized exoplanets are predominantly rocky: these two groups merge at about 2–3 Earth radii.

Giant planets have been discovered in the circumstellar habitable zones of some exoplanetary systems, suggesting that life might be possible on moons in orbit around them. If Jupiter, for example, were to migrate into the habitable zone of the Sun, we might expect surface oceans to be present on Europa, Ganymede and Callisto. However, with reference to figure 4.4, it is clear that raising the mean surface temperature to ~290 K on such worlds would lead to a steady loss of volatiles, as H_2O evaporating from the oceans would be subject to leakage into space. Hence, they would likely evolve to become dry and airless, perhaps resembling the Earth's Moon: a mass at least as great as that of Mars would be needed to avoid this fate. Moreover, the surface environments of Jupiter's moons are exposed to energetic solar-wind particles streamed by the planet's strong magnetic field, a lethal hazard to surface life. If giant exoplanets possess similar magnetic fields the same problem would arise.

Another important comparison is shown in figure 9.7, which plots orbital eccentricity with respect to semi-major axis. The orbits of the major planets in our local system are nearly circular ($e < 0.1$ for all except Mercury). In the exoplanet

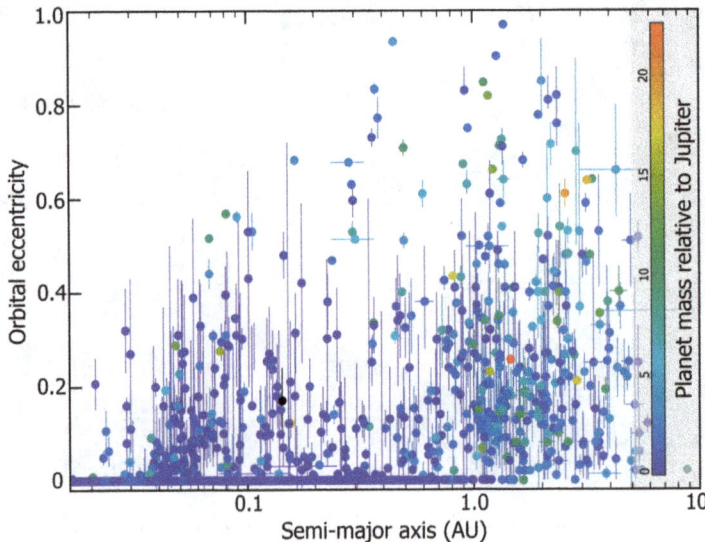

Figure 9.7. Distribution of orbital eccentricity versus semi-major axis for exoplanets. Points are color-coded by planetary mass relative to Jupiter, according to the scale on the right. For comparison, in our solar system only one of the eight major planets (Mercury, $e = 0.21$) has an eccentricity greater than 0.1. Based on data from the Exoplanet Orbit Database and Exoplanet Data Explorer at exoplanets.org (Han *et al* 2014).

population, however, there are many examples with much higher values. This finding presents a challenge in understanding their origin—a uniformly-rotating protoplanetary disk should generally result in reasonably circular orbits; perhaps disruption by a neighboring star is implicated. It also has major consequences for habitability. An earthlike planet in a highly eccentric orbit would experience extreme 'seasons' if it spends only part of the year in the habitable zone: a possible scenario is shown in figure 9.8. However, a far greater hazard to habitability arises in a system where *giant* planets occupy such orbits. If, say, Jupiter in our solar system were perturbed into an orbit similar to that shown in figure 9.8, the Earth and the other inner planets would be very rapidly dispersed or destroyed. Many exoplanets with highly eccentric orbits are of Jupiter mass and above, and we can probably discount their systems as potential hosts for life. Similarly, systems containing hot Jupiters, even those in nearly circular orbits, also seem unpromising. Such planets probably did not originate at their present locations close to the star: it is much more likely that they formed at greater distances in the protoplanetary disk and subsequently migrated inwards, thereby dispersing any smaller planets that had formed inside their original orbits.

9.1.5 Super-earths

Super-earths—predominantly rocky planets significantly larger and more massive than the Earth—are evidently ubiquitous in the exoplanet population (figure 9.6). As we have no local examples to explore, we must develop models to elucidate the nature of these worlds, extrapolating from those developed to describe planets in our solar system. Let us consider two examples, each 50% larger than the Earth but

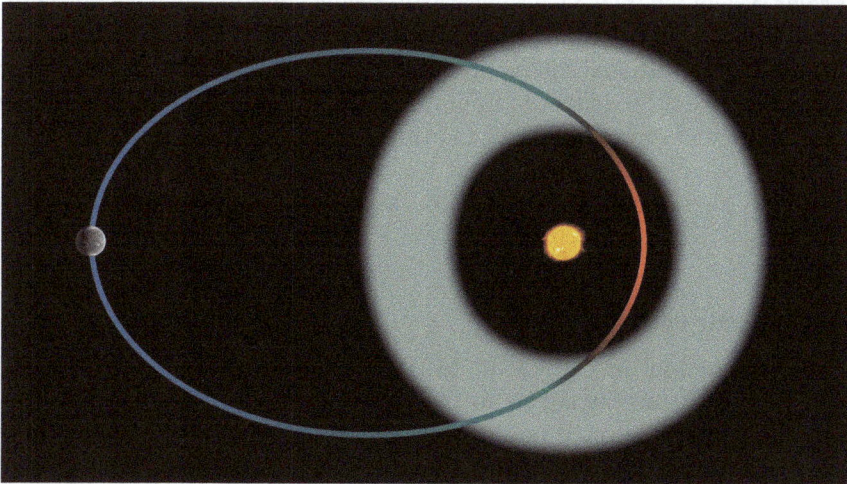

Figure 9.8. Schematic example of a hypothetical exoplanet in an eccentric orbit ($e = 0.75$) that carries it into and out of the circumstellar habitable zone (green annulus) as it follows its path around the star. Orbital speed is least when the planet is farthest from the star, so it spends a large proportion of time beyond the cold outer boundary of the habitable zone, and a much shorter time inside its hot inner boundary (blue and red segments, respectively).

differing in locus of formation with respect to the frost line of the protoplanetary disk (section 4.2.2).

Silicate super-earth: First consider a planet formed inside the frost line of the system, composed primarily of silicate rock and metals in similar proportions to the Earth and with similar mean bulk density: the greater volume thus corresponds to a factor ~3.4 increase in mass relative to Earth. Models predict an internal structure resembling the standard core-mantle-crust model for the Earth (figure 5.3), but lacking a liquid outer core because of the greater internal pressure; this being the case, the planet will lack a significant magnetic field. The Earth's magnetic field contributes to retention of the atmosphere by shielding it from erosion by solar wind particles; however, the greater escape velocity (~17 km s^{-1} in this example, compared with 11.2 km s^{-1} for Earth) should ensure adequate retention. Geological activity is likely to match or exceed that of the Earth in both intensity and longevity, resulting in copious, long-term release of atmospheric gases by volcanism and tectonic recycling (section 5.3). Free H may be retained somewhat more efficiently than on Earth (see figure 4.4), so the atmosphere may be more reducing and hence more conducive to organic synthesis (section 1.4.1). If located in the circumstellar habitable zone of its parent star and adequately endowed with water and organic carbon, such a planet would appear to be an excellent candidate for life.

Ocean-world super-earth: Such a planet forming beyond the frost line of the system would acquire large quantities of ice as well as silicates and metals. In this case, the internal structure may be qualitatively similar to that of a large icy moon such as Ganymede (figure 8.4), with a metallic core and silicate mantle beneath a subsurface ocean and an icy crust. However, unlike Ganymede and other examples in our solar system (chapter 8), a super-earth-sized icy world may generate sufficient internal heat to maintain a liquid subsurface ocean without reliance on tidal friction. In some cases, gravitational interactions with other planets or with the protoplanetary disk may cause the planet to migrate inward to occupy an orbit closer to the star than its original formation zone, and this may lead to melting of the icy crust to produce a global *surface ocean*. Given the prediction of large volumes of liquid water, either at or below the surface, such planets are evidently potential hosts for life.

9.1.6 The galactic habitable zone

The Sun is situated in the disk of the Milky Way, at a radial distance about two-thirds of the way from the center to the edge (figure 9.9). The large majority of exoplanetary systems detected to date are similarly located, within a few thousand light years of the Sun (i.e. within about a tenth of the distance from the Sun to the galactic center). Does this location affect habitability? There are several reasons to suppose that it does, leading to the concept of a galactic habitable zone (Lineweaver *et al* 2004). The number of stars per unit volume increases rapidly toward the galactic nucleus and declines toward the edge. This has two consequences: (i) production and distribution of the chemical elements needed for life occurs

Figure 9.9. Schematic representation of our Galaxy, the Milky Way, indicating the locus of the Sun and the extent of the proposed galactic habitable zone (green ring). Regions inside the inner edge of the ring are more liable to disruption by supernova explosions, regions beyond its outer edge are deficient in the necessary chemical elements. Image credit: NASA/Caltech.

most rapidly toward the nucleus and relatively slowly toward the edge; and (ii) the potential for extant life to be destroyed by a catastrophic event, such as a nearby supernova explosion, is greatest toward the nucleus and least toward the edge. Thus, an intermediate locus is favored as a compromise between these opposing trends, as illustrated in figure 9.9. A further consideration is the locus of the system with respect to the spiral arms of the Galaxy. The Sun's present position between spiral arms is favorable, as it reduces the possibility of potentially disruptive encounters with dense interstellar clouds that might disturb the Oort cloud and result in cometary collisions with the planets.

9.2 The search for spectroscopic biosignatures

9.2.1 The Earth as a template

In December 1990, the Galileo spacecraft made extensive observations of an earthlike planet to search for life. The results were positive, including detections of atmospheric biosignatures (oxygen and methane in extreme disequilibrium) and surface biosignatures (widely distributed surface pigments with a sharp absorption edge in the red part of the visible spectrum), as well as radio signals uniquely attributed to intelligence (Sagan *et al* 1993). The planet was, of course, the Earth itself, and the observations were made by a mission destined to explore Jupiter and its moons. Nevertheless, this investigation was a landmark demonstration of principle, illustrating how we might conduct future searches for biosignatures on exoplanets—yet to be discovered at the time that it was carried out—using the Earth as a template.

The technical challenges involved in applying equivalent spectroscopic methods to studies of exoplanets are inevitably very much greater. Nevertheless, progress has been made. In the case of a transiting exoplanet (figure 9.1), observations allow the composition of its atmosphere to be investigated in a manner analogous to studies of Venus during solar transit: absorption features are produced as starlight filters through the atmosphere around the limb of the planet's disk as it passes in front of the star. Half an orbital period later the exoplanet goes into eclipse behind the star, and this allows the possibility to detect light from the full disk of the planet by comparing the radiation received just before or after the eclipse (star + planet) with that received during the eclipse (star alone). These methods have been applied successfully to studies of giant exoplanets. With the advance in sensitivity enabled by facilities such as the James Webb Space Telescope, it will be feasible to study earth- and super-earth-sized planets in the habitable zones of transiting systems in the same way. Other techniques under consideration for future missions include coronagraphy, a development of methods used to study the solar atmosphere by masking the disk of the Sun, and interferometry, which would enable the signals from star and planet to be separated by interference of radiation collected by multiple telescopes.

9.2.2 Atmospheric methane and oxygen

Atmospheric methane (CH_4) is a potential biosignature, as discussed in the context of Mars in section 7.2.3. Observed in isolation, CH_4 does not provide a definitive result because it can emanate from non-biological sources, such as geochemical processes or volcanism on a rocky planet, and cryovolcanic activity on an icy world such as Titan. Molecular oxygen (O_2 or O_3) is more promising, as photosynthesis is the only known source capable of producing the levels of free oxygen found on Earth (photodissociation of H_2O may also contribute in some circumstances, but at a lower level). Best of all would be simultaneous detection of both CH_4 and O_2 in the same atmosphere. These molecules are said to be in disequilibrium because they readily react with each other (to form CO_2 and H_2O), and the reaction continues until the least abundant of the two is consumed unless it is replenished by an active source. On Earth, O_2 is plentiful and CH_4 is present as a minor constituent, emanating primarily from methanogenic life. Detection of these molecules together in an exoplanetary atmosphere would make a compelling case for extant life—the holy grail of atmospheric biosignatures.

Simulations illustrating examples of spectra that might be obtained by the eclipse method described above appear in figure 9.10. Because any planet in the circumstellar habitable zone of a star must have global temperatures within a limited range,

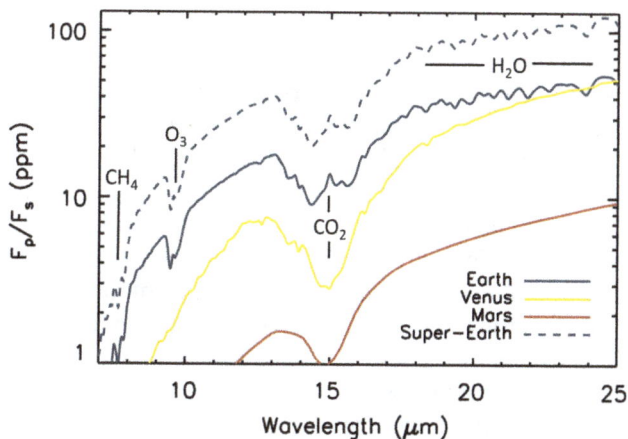

Figure 9.10. Model simulations of infrared spectra for planets in orbit around a red dwarf star of surface temperature 2800 K. The vertical axis is the planet/star flux ratio in parts per million. Planets identical to the present-day Earth (solid blue), Venus (yellow) and Mars (red) are shown, together with a super-earth (dashed blue), 50% larger but with the same atmospheric composition as the modern Earth. Image credit: Tyler Robinson.

such planets are always expected to be brightest in the mid-infrared region of the spectrum (see figure 9.5, blue curve), so this is an appropriate choice for observation. Spectral absorption features (dips) produced by CH_4, O_3, CO_2 and H_2O are labelled in figure 9.10. Note that O_2 cannot be detected directly at these wavelengths as it lacks absorption features in the infrared, but O_3 (ozone) has an absorption feature at 9.5 μm. O_3 is formed in the Earth's atmosphere by a two-step process of O_2 photodissociation and $O_2 + O$ recombination, and this can occur only in an atmosphere that already contains abundant O_2. Because of this, we may use O_3 as a proxy for O_2.

Another point that must be taken into account is that planetary atmospheres evolve with time. Biosignatures may not build up to measurable levels until some considerable time after the origin of the species that produces them. A prime example is oxygen in the Earth's atmosphere, levels of which remained low for at least a billion years after the origin of photosynthetic life (section 6.2.2). The rise in atmospheric O_2 was delayed primarily by oxidation reactions that continued until surface elements such as Fe were fully oxidized. Because of this, no trace of O_3 would be detectable in the spectrum of the Archean Earth, and exoplanets at a similar stage of development may also lack this key biosignature.

9.2.3 The red edge

Observations of earthlike exoplanets in the visible to near infrared region of the spectrum may also be enabled by future generations of optical telescopes. At these wavelengths, the radiation received from the planet is reflected starlight rather than the planet's own thermal emission, but it may nevertheless bear the spectroscopic fingerprints of the planet's surface and atmosphere. For example, observations in this spectral region provide an opportunity for a direct detection of atmospheric O_2. Reflectance spectroscopy is a well-developed and powerful technique used to study surface materials on Earth and other solid bodies in the solar system. Extending it to potentially habitable exoplanets opens up an intriguing possibility to detect surface biosignatures, such as the absorbers that drive photosynthesis.

Photosynthetic harvesting of energy from our home star is a vital process for life as we know it, and it seems reasonable to assume that life elsewhere has developed an equivalent mechanism. Terrestrial photosynthetic organisms—green plants, algae and cyanobacteria—contain pigments that are efficient absorbers of the Sun's radiation, thereby gathering the energy needed to trigger endothermic reactions that convert CO_2 into glucose. These pigments have characteristic spectra that can be readily identified by remote observation, given data of sufficient quality. This technique is used to study the Earth from space by facilities such as NASA's Earth Observing System, which maps the distribution of photosynthetic organisms both on land and in the oceans. Chlorophylls absorb strongly in the blue and red regions of the visible spectrum but reflect other wavelengths (figure 9.11), thus accounting for the green color of plant life. Other compounds contribute to the absorption of sunlight, including carotenes, which absorb strongly in the blue and green but reflect in the red. The most striking feature seen in photosynthetic biomass containing a typical mix of these pigments is a rapid decline in absorbance at the red end of the visible spectrum for wavelengths longer than 700 nm. This 'red edge' is therefore a potentially important biosignature.

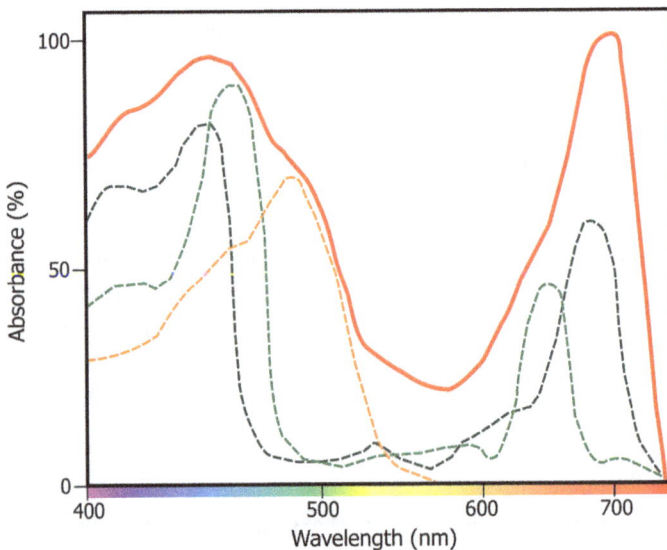

Figure 9.11. Absorbance spectra of terrestrial leaf pigments in the visible (dashed curves): chlorophyll a (dark green), chlorophyll b (light green), and beta carotene (orange); the solid red curve shows a typical spectrum of vegetation containing these pigments. The steep decline in absorbance at wavelengths longer than 700 nm (the red edge) is considered a particularly sensitive biosignature.

It should be noted that the pigments adopted by terrestrial biology have spectra that are well matched both to the Sun's radiation (figure 9.5, yellow curve) and to the transparency of the Earth's atmosphere to visible light. Conditions on other planets orbiting other stars might necessitate different choices. For example, the peak in stellar brightness occurs at wavelengths significantly beyond the red edge for the coolest red dwarfs (figure 9.5), and the nature of the absorbers that photosynthetic life might adopt on a planet in such a system is unknown. The best candidate for a detection of the red edge would thus appear to be an exoplanet in which atmospheric oxygen has already been detected, orbiting in the habitable zone of a star of surface temperature not dramatically different from that of the Sun (say, 4500–6000 K), thus maximizing the probability that a similar set of pigments is being utilized. However, the sensitivity needed for such an investigation is beyond current feasibility limits.

9.3 The search for extraterrestrial intelligence (SETI)

9.3.1 Rationale and brief history

Spectroscopic biosignatures such as those discussed in the previous section might yield a detection of life beyond our solar system, but they are unlikely to distinguish a planet hosting only microbial life (such as methanogens and photosynthetic bacteria) from one hosting more evolved life forms. Detection of life that has reached or exceeded the evolutionary development of our own species requires a different approach[2]. The advent of radio communications technology within the last 100 years has given our species the opportunity to search for signals from extraterrestrials possessing similar technology; it might also (inadvertently) reveal our presence to them.

The first systematic attempt to search for radio signals from extraterrestrial life was the Project Ozma experiment, led by Frank Drake, which began in 1960, more than three decades before the first exoplanet detections were made. Two nearby main-sequence stars thought likely to host planetary systems were observed, and a search was made for signals that could not be explained by natural phenomena but might originate as narrow-band radio transmissions typical of those used in terrestrial communication. Subsequent work has steadily extended both the scope and sensitivity of the search. Modern facilities such as the Allen Telescope Array, equipped with state-of-the-art receivers and supported by the latest data handling and analysis software, permit searches many orders of magnitude more sensitive than the original Project Ozma. Observations have been made that focus on individual targets, such as stars with known planetary systems; others spread the net more widely, surveying large areas of the Milky Way and entire galaxies. The spectral range has also been extended to include laser signals at visible wavelengths, as a potential alternative to radio signals as a means of interstellar communication. To date, none of these observations has resulted in a confirmed positive detection.

---

[2] It is presumed that pollutants, such as fluorocarbons released to the Earth's atmosphere by human activity, do not reach levels sufficient to become observable biosignatures!

9.3.2 The Drake equation

The Drake equation (figure 9.12), formulated by Frank Drake in 1961, attempts to provide a framework for estimating the number (N) of civilizations in our Galaxy with which interstellar communication is possible. Unlike other famous equations, such as Einstein's $E = mc^2$, the Drake equation does not describe any mathematical or physical law; neither is it an empirical law in a meaningful sense, as some of the terms on which the solution depends are unconstrained by available data and may remain so in the foreseeable future. Instead, the Drake equation is best regarded as a holistic *roadmap* for SETI (Cabrol 2016); as such, it serves as a fertile stimulant for ideas, discussion and informed speculation.

The first three terms on the right-hand side of the equation are reasonably well constrained by observational data. The star formation rate in our Galaxy is estimated to be in the range 5–10 stars per year. The fraction of stars that have planets is now known to be quite high, at least amongst single stars, so $f_p \sim 0.6$ would be a reasonable general estimate. The average number of planets per planetary system that have the potential to develop an ecosystem (i.e. those of appropriate size that lie in a habitable zone) is also constrained by the latest research, suggesting $N_e \sim 0.4$. The product of these numbers leads to an estimate that our Galaxy is producing an average of about two new planets per year that are potential hosts to life. Considered over cosmic timescales, this rate implies that such planets exist in enormous numbers: if the rate were to remain constant for, say, a tenth of the age of the Galaxy, the number of habitable planets produced in that time would be about 3 billion.

We must remember, however, that 'habitable' and 'inhabited' are not the same. The temptation to conclude from such numbers that the cosmos is teeming with life must be resisted, as the other terms in the equation are unknown and might be extremely small. The probability that life develops under favorable conditions (represented by the term f_l) may be clarified by future research, as we seek to understand the processes that led to life's origin on Earth (chapter 6) and to detect biosignatures on other planets (section 9.2). At present, the only significant

Figure 9.12. The Drake equation, illustrated with a definition of each term. Image credits: Danielle Futselaar and the SETI Institute.

constraint on f_l is the finding that life on Earth appeared quite early in the history of our planet (section 1.3), which provides the merest hint that f_l might be close to unity. If life turns out to be common, how likely is it to develop intelligence (f_i) and the capacity for interstellar communication (f_c)? And if such civilizations exist, what is their average lifetime (L)? Here we enter the realms of speculation. We cannot presume that evolution is bound to produce intelligence, and we cannot predict the nature of other technological civilizations or how long they might exist. Any value of N is therefore possible, subject only to the limit $N \geqslant 1$ imposed by our own existence.

9.3.3 Where is everyone?

The technological development of our species has entered a period of exponential growth. From laptops to supercomputers, from smart phones to communication satellites, from electron microscopes to space telescopes, our technology has evolved in ways it would have been hard to imagine just a century ago, and it shows no signs of slowing down. The timescale over which this has occurred is infinitesimal in comparison to the age of the Galaxy. Whether such growth is sustainable and where it might ultimately lead are interesting questions to consider, but let us assume that other civilizations undergo similar transitions, rapidly reaching levels of sophistication that surpass our own. It follows that any civilization that we might encounter as the result of a successful SETI observation is likely to be technologically more advanced than us: the probability of meeting our peers, i.e. civilizations at a closely similar stage in such rapid development, is extremely small.

In the light of this prediction, the continuing absence of a confirmed detection of extraterrestrial intelligence appears significant and perplexing. An argument attributed to Enrico Fermi (the Fermi paradox; see Livio and Silk 2017) postulates that if advanced life exists we should have encountered it by now. The possibilities for contact are not necessarily limited to radio waves or other electromagnetic signals that might be detected by astronomical observation. Advanced civilizations are presumed to have overcome the challenges of interstellar travel, in which case their spacecraft (robotic or otherwise) could permeate the Galaxy on timescales that are short by the standards of cosmic evolution. Yet no convincing evidence of visitations to our solar system by aliens or their artefacts has been found.

Several interpretations of the Fermi paradox may be considered:

- The origin of life and/or the emergence of advanced life are extremely improbable events, perhaps even unique to our planet (the rare Earth hypothesis); some or all of Drake's parameters f_l, f_i and f_c may be infinitesimally small.
- Technological development is inevitably accompanied by a tendency to self-destruct, e.g. by nuclear war or environmental catastrophe (Drake's parameter L is short).
- Advanced civilizations are in fact common; perhaps they choose not to advertise their presence, or perhaps our strategies for detecting them are incorrect or lack the necessary sensitivity.

A criticism of the rare Earth hypothesis is that it designates undue status to our species and our planet. The history of scientific discovery contains many examples of hypotheses that have falsely assigned us to a special place in the Universe (see Sagan and Newman (1983) for an eloquent summary). We now know that we live on an unexceptional planet orbiting an unexceptional star in an unexceptional region of an unexceptional galaxy. The proposal that our planet is the only one to host intelligent life should therefore be tabled until all reasonable steps have been taken to prove otherwise. A related idea is the possibility that life is common, but the average time for intelligence to evolve on a life-bearing planet is much longer than was the case on Earth: perhaps by serendipity we were amongst the earliest to appear, with a galaxy-wide emergence of advanced civilizations still in the future.

If extant civilizations do currently exist and are, indeed, more advanced than us, this leads to another dilemma: would it be prudent for us to avoid advertising our presence to them? The history of our own species suggests that caution may be the best policy. Colonization of other continents by European nations, to the general detriment of indigenous inhabitants, may be cited as symptomatic of the possible consequences. In fact, it may already be too late to take such a decision. Our species has been transmitting radio signals for more than a century, with no particular concern to prevent their leakage into space. An advanced civilization within 100 light years of Earth with a powerful enough radio telescope could already be listening.

9.4 What if we succeed?

I conclude this chapter with a brief assessment of some possible consequences arising from a detection of extraterrestrial life. The nature of the outcome will doubtless depend on the nature of the discovery, the possibilities ranging from identification of microbial life within or beyond our solar system to contact with an advanced civilization.

9.4.1 Dissemination

On August 6, 1996, it was announced that evidence had been found for fossil bacteria in a meteorite from Mars. This news made headlines around the world and prompted the then president of the United States, Bill Clinton, to make a formal televised announcement. An uncredited source quoted by the Cable News Network (CNN) described it as 'arguably the biggest discovery in the history of science.' Few might disagree with this assessment if the evidence had proven to be conclusive. However, it soon became clear that the evidence, although not necessarily incorrect, was open to other, less dramatic interpretations (see section 7.2.2), and the story soon disappeared from the news media. It nevertheless provides an interesting case study that gives insight into the short-term impact of such an announcement, including possible spin-off effects that, on this occasion, included a renewed interest in the search for extraterrestrial life on Mars and elsewhere. It also emphasizes the importance of rigorous, dispassionate assessment of the evidence and full consideration of all alternative interpretations. Extraordinary claims demand extraordinary proof.

In general, scientists have both an obligation and a disposition to disseminate the results of their research, especially those of greatest potential significance. This is, indeed, exemplified by the Martian meteorite event, where the press release was linked to contemporary publication of the research in a respected scientific journal. It seems reasonable to suppose that a similar protocol would be followed if, say, evidence is found for atmospheric biosignatures on an exoplanet. Would a discovery of extraterrestrial intelligence be treated in the same way? Assuming that the detection is the result of an astronomical observation, some initial sharing of information amongst the scientific community would be essential to enable corroboration by other observers using other facilities, and to allow a consensus to be reached on the correct attribution of the signal. The scientific community has drawn up a set of protocols to guide the discoverer through this process[3], and has resolved that the discovery, once confirmed beyond reasonable doubt, should be disseminated promptly, openly, and widely through both scientific channels and public media. However, these protocols are not legally binding and there would presumably be political as well as scientific issues to consider that might affect the timing, format and content of a public announcement.

9.4.2 Implications

If life (presumably, microbial life) is found elsewhere in our solar system, perhaps on Mars or in the cryovolcanic plumes of an icy moon, the first question to address would be rather mundane: have we merely rediscovered terrestrial biology? Did we inadvertently contaminate other worlds in our search for indigenous life? This possibility might be ruled out by context (e.g. fossilized Martian life would clearly predate any modern contamination) or by phylogenetic analysis in the case of extant life. Another potential source of contamination, more interesting and harder to exclude, is impact-generated exchange of materials that carried life from one world to another at some time in the past: can we be sure that our discovery is an independent instance of an origin of life, or is it another manifestation of the same origin? In either case, if extant life is found, extreme care will be needed to quarantine samples that may be returned to Earth (either intentionally or inadvertently) to avoid the risk of any pathogenic effect on our biosphere. Phylogenetic comparisons between terrestrial microbial life and an independent instance from another world would be the ultimate scientific goal. Then the text books would need to be rewritten.

Secure detections of biosignatures on exoplanets should not be affected by such ambiguity: barring the remote possibility of interstellar panspermia, they would surely represent independent instances of an origin of life. In purely practical terms, the scientific impact would be more limited than that arising from a detection of life in our solar system. No morphological or phylogenetic comparisons would be

---

[3] See https://www.seti.org/post-detection.html.

possible, and our knowledge of the environment in which life arose would be limited to global parameters such as mean temperature and the spectrum of incident light from the parent star. Beyond the 'wow factor' of such a discovery, the main outcome would be a revised assessment of the probability that life arises when the conditions are favorable. We would know that Drake's f_l parameter cannot be extremely small.

The consequences of discovering extraterrestrial intelligence are, of course, much harder to predict, but some contingency plans can be made. Following an announcement, much discussion and speculation would doubtless ensue concerning the nature and intent of the aliens, and how to decipher their signals. The expertise required to address such questions would extend beyond the scientific disciplines normally associated with research in astrobiology: see, for example, Blumberg (2011), Denning (2011), Harrison (2011) and Cabrol (2016) for some interesting perspectives. Under the auspices of the International Academy of Astronautics, the scientific community has drawn up a set of protocols to address the question of whether signals should be transmitted by us in response to signals received[4]. It is recommended that this decision should be taken by the United Nations General Assembly, after extensive, international consultations with all interested parties, to ensure that both the choice to transmit and the content of any such message should reflect a consensus on behalf of our species. However, as before, these protocols are not legally binding.

Let us conclude on an optimistic note. If communication is established with an altruistic alien civilization, we may learn much from the exchange—about ourselves as well as about life in the Universe. It would be reassuring to know that Drake's parameter L is not too terribly short.

Questions and discussion topics

- Suppose that alien astronomers from two different planetary systems in our Galaxy are observing our Sun, each using powerful telescopes and sensitive instruments that can detect exoplanets by the transit and Doppler methods. By chance, one of them is situated in a direction parallel to the ecliptic plane (the plane of the Earth's orbit around the Sun), the other in a direction perpendicular to it. What do you expect each of them to see?
- Consider why planetary magnetic fields are thought to be generally favorable to habitable conditions on earthlike planets, but may be detrimental to habitable conditions on the moons of giant planets.
- Would you expect chlorophyll-bearing plants to be capable of photosynthesizing on an exoplanet in the circumstellar habitable zone of a star significantly cooler than the Sun, such as a red dwarf with a surface temperature <3000 K?
- Why is CO_2, a product of metabolism, not considered to be a useful atmospheric biosignature?

---

[4] See http://www.setileague.org/iaaseti/protocol.htm.

- Use the Drake equation (figure 9.12) to make your own estimate of N. Ask your friends or classmates to do the same and compare the results.
- Consider whether there is any reason to suppose that SETI observations directed at planetary systems close to the ecliptic plane in the sky are more likely to succeed than those directed elsewhere.
- Consider how you would vote if a planet-wide referendum was held to decide whether to respond to a signal from an alien civilization. Assume that nothing is known about the nature of the civilization other than its location, on a small planet somewhere in the vicinity of Betelgeuse.
- Should SETI be a major focus for research effort and resources, or is prioritizing the search for more primitive forms of extraterrestrial life a better strategy?

References and further reading

Blumberg B S 2011 Astrobiology, space and the future age of discovery *Phil. Trans. R. Soc.* A **369** 508

Cabrol N A 2016 Alien mindscapes: a perspective on the search for extraterrestrial intelligence *Astrobiology* **16** 661

Campante T L *et al* 2015 An ancient extrasolar system with five sub-earth-size planets *Astrophys. J.* **799** 170

Denning K 2011 Is life what we make of it? *Phil. Trans. R. Soc.* A **369** 669

Drake F 2013 Reflections on the equation *Int. J. Astrobiol.* **12** 173

Han E *et al* 2014 Exoplanet orbit database II. updates to exoplanets.org *Pub. Astron. Soc. Pacific* **126** 827

Harrison A A 2011 Fear, pandemonium, equanimity and delight: human responses to extraterrestrial life *Phil. Trans. R. Soc.* A **369** 656

Kaltenegger L 2017 How to characterize habitable worlds and signs of life *Ann. Rev. Astron. Astrophys.* **55** 433

Korpela E J 2012 SETI@home, BOINC, and volunteer distributed computing *Ann. Rev. Earth Planet. Sci* **40** 69

Lineweaver C H, Fenner Y and Gibson B K 2004 The galactic habitable zone and the age distribution of complex life in the milky way *Science* **303** 59

Livio M and Silk J 2017 Where are they? *Phys. Today* **70** 50

Sagan C and Newman W T 1983 The solipsist approach to extraterrestrial intelligence *Quart. J. R. Astron. Soc.* **24** 113

Sagan C, Thompson W R, Carlson R, Gurnett D and Hord C 1993 A search for life on earth from the galileo spacecraft *Nature* **365** 715

Tarter J 2001 The Search for extraterrestrial intelligence (SETI) *Ann. Rev. Astron. Astrophys.* **39** 511

Wright J T and Gaudi B S 2013 *Exoplanet detection methods Planets, Stars and Stellar Systems vol. 3: Solar and Stellar Planetary Systems* ed T Oswalt, L French and P Kalas (Dordrecht: Springer) p 489 http://dx.doi.org/10.1007/978-94-007-5606-9_10

www.ingramcontent.com/pod-product-compliance
Lightning Source LLC
Chambersburg PA
CBHW081540220326
41598CB00036B/6498